Estimator's Piping

MAN-HOUR TOOL

Second Edition

Gustavo Cinca

Estimating Man-Hours for Carbon Steel Process Piping Projects: A Practical Manual with Examples

First Edition: August 2018

Second Revised and Expanded Edition: August 2026

Published by: Gustavo Cinca

COMPANION DIGITAL RESOURCE FOR READERS

PIPING MAN HOURS TOOL (PMHT)

Multi-Standard Web & Mobile Estimation Application (ASME B31.3 · DIN/EN · GB/CN)

Dear Reader,

To complement the formulas, benchmark tables, and labor productivity methodologies presented in this book, author **Gustavo Cinca** provides direct reader access to the official **Piping Man Hours Tool (PMHT)**. This mobile and cloud-based engineering application calculates direct man-hours, incorporating field multipliers and environmental corrections based on selected parameters. It complies with international piping standards(**(ASME B31.3, European DIN/EN, and Chinese GB/CN) and runs seamlessly across any device (PC, Mac, iPad, iPhone, and Android)**

HOW TO ACCESS & USE THE COMPANION APPLICATION

Steps to activate your reader access:

1. Scan the QR code or navigate to:
   `https://calculatemanhours.com/activate/`
2. Enter your email to receive your authorized reader license.
3. Launch the application directly in your browser or install as PWA:
   `https://calculatemanhours.com/pmht-tool/`

Scan for Instant Access

Multi-Standard Coverage & Productivity Multipliers:

- ASME NPS 1/2"–24" • DIN/EN (DN 15–600 / PN) • GB/CN (DN / MPa)
- MTO Excel Importer • Bare & Water Loads • Prefabrication Degree
- Brownfield / Congestion • Work Elevation • Extreme Weather

ADDITIONAL RESOURCES & LIFETIME UPGRADES

By activating their account, readers gain access to continuous updates for international databases (ASME, DIN/EN, and GB/CN), the ability to directly import MTO spreadsheets, and executive reports in PDF and Excel formats.

Reader Activation Portal:

`https://calculatemanhours.com/activate/`

Technical Inquiries & Support: `support@calculatemanhours.com`

`https://calculatemanhours.com`

Contents

Prologue

Accurately estimating man-hour consumption for project execution is essential for the smooth operation of any company. This document provides a reliable tool for estimating man-hours in the construction industry, with a focus on installing overhead carbon steel piping.

The Importance of Accurate Estimation

When a company submits an extremely low bid and is awarded the contract, initial joy can quickly turn to panic when faced with imminent financial problems. The execution of the project will be impacted to a degree proportional to the deviation of the estimate. In the best-case scenario, there will be schedule delays. In the worst-case scenario, the company could face bankruptcy. Either way, the client is also harmed, and the company's valuable market reputation is damaged.

Conversely, if a company's bid is too high compared to its competitors', it will likely lose the contract, which could damage the commercial relationship with the client. Ultimately, this represents a lost business opportunity, which reduces revenue and increases overhead costs for the bidder.

Purpose of This Publication

Accurate bidding is essential for the profitable execution of contracts. This publication aims to provide beginners, contractors, estimators, project controllers, and shareholders with a reliable tool estimating man-hours accurately..

Contents of the Manual

This manual explains how to calculate direct man-hour consumption for installing overhead carbon steel piping designed in accordance with the ASME B31.3 standard.

Chapter 1

Introduction

1.1 General Considerations

This book presents a calculation procedure for estimating the direct man-hours required for the **above-ground installation of process piping**, including its load-bearing supports and corrosion protection.

The reader will find an extensive database of man-hour consumption for various construction activities. This data, which has been field-verified by the author, is indispensable for preparing accurate budgets.

First, the conditions under which a task is completed must be defined in order to measure the man-hours needed to perform that task.

Standardizing these conditions guarantees objective and repeatable measurements. This set of criteria is referred to as the **Standard Conditions**.

1.2 Standard Measurement Conditions

The man-hour (Mhr) rates per joint detailed in this manual apply specifically to the **Shielded Metal Arc Welding (SMAW)** process.

These reference values are based on the fourteen standard working conditions described below:

Location The work is performed at a standalone site, separate from any operating plant.

Typical Crew The indicated rate corresponds to a crew composed of two journeymen and one helper. The welder's qualifications will depend on the type of joint.

Climate The tasks are performed in a moderate climate, with average temperatures between 41 °F to 86 °F (5 °C to 30 °C), and winds below 18.6 mph (30 km/h).

Contaminants The work area is free of environmental contaminants.

Schedule Work is performed during daytime hours.

Terrain The project is developed on level and dry ground.

Working Height The maximum height of the working plane does not exceed 5 feet (1.50 meters) above ground level.

Clearance In roofed areas, the minimum overhead clearance is 8.5 feet (2.60 meters).

Material Transport The maximum distance from the storage area to the installation site is 985 feet (300 meters).

Personnel Access The personnel access gate is 650 feet (200 meters) or less from the work area.

Engineering and Supplies Engineering deliverables and materials are received on time, according to the project schedule.

Personnel Tasks are executed by properly trained field personnel.

Equipment Construction equipment and tools appropriate for each activity are available.

Altitude The worksite is at an altitude that does not require additional physical exertion from workers.

1.3 Scope and Exclusions

The man-hour calculations presented in the tables do **not** include the following items:

- ➔ Scaffolding assembly and dismantling.
- ➔ Rework hours due to design or fabrication errors.
- ➔ Labor for the initial adjustment or re-torquing of bolts in flanged connections.
- ➔ Fabrication of special fittings.
- ➔ Installation of thermal insulation on piping.

Chapter 2

Estimating Man-Hours for Piping Supports

Supports are the elements that transfer loads from a piping line to a support structure or a piece of equipment, and are regulated by standards such as ASME B31.3 and MSS SP-58.

For the purposes of this manual, the term *support* refers only to material welded directly to the pipe. Otherwise, it would be impossible to establish a general estimation model.

The location and characteristics of supports depend on multiple factors:

- ➔ Pipe diameter and wall thickness.
- ➔ Fluid density and temperature.
- ➔ Line configuration and routing.
- ➔ The presence of concentrated loads (valves, meters, etc.).
- ➔ Type of available support structure (sleepers, gantries, foundations, etc.).
- ➔ Presence of thermal insulation.

Since estimators generally lack detailed engineering information about the supports during the bidding stage, an estimation method is necessary to quantify them.

2.1 Support Estimation Method

Table 9.1 provides labor rates for straight and simple pipe runs. To use the table for an actual piping that includes fittings, valves, and flanges, first convert its complexity into an **"equivalent length"**. This calculation allows us to standardize the estimation process.

The process consists of three simple steps:

1. **Calculate the Total Weight.** First, sum the individual weights of all components in the run (e.g., pipes, elbows, flanges, valves) to obtain the total mass of the system.
2. **Determine the Equivalent Length.** Next, divide the total weight (calculated in Step 1) by the unit weight (in lb/ft or kg/m) of the main pipe diameter of the run. The result is a theoretical length representing the entire run as a single straight pipe.
3. **Estimate the Man-Hours (Mhr).** Using the main diameter and the equivalent length from Step 2, refer to Table 9.1. Obtain a rate (Mhr/ft or Mhr/m) from there, which, when multiplied by the equivalent length, yields the total man-hours for the run.

2.2 Scope of Support Calculation

➔ The man-hours for fabrication and installation included in the tables only cover elements that are directly attached to the pipe, such as wear plates, lugs, and pipe shoes. This excludes anchors and fixed points, whose design and installation depend on a dedicated piping stress analysis.

➔ The cleaning and painting of these elements are calculated according to the methodology in Section 10.2 and Table 10.2 of this manual.

➔ Structural elements, such as hanger rods, spring supports, and large support structures (e.g., pipe racks and trestles), are not included.

2.2.1 Support types whose fabrication and assembly times are included in Table 9.1

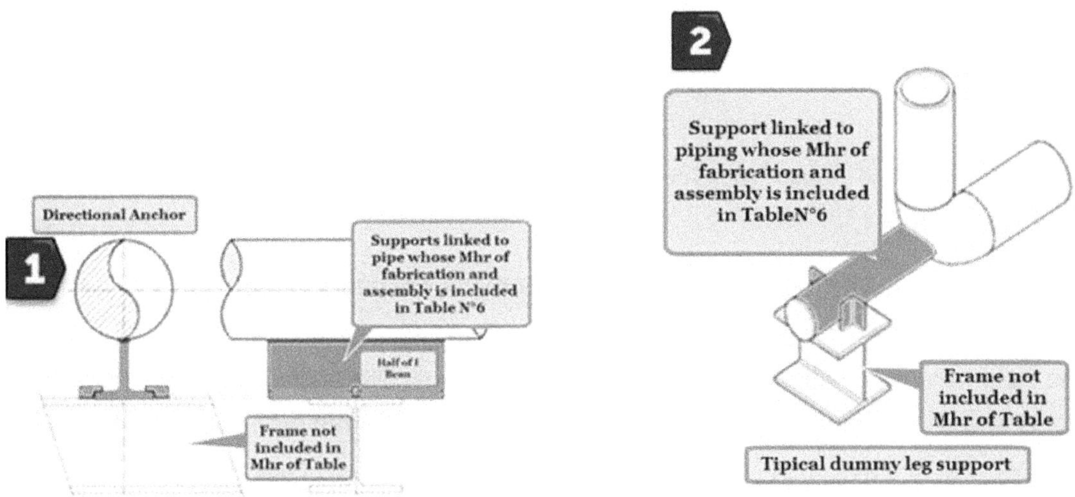

Figure 2.1. Types of supports INCLUDED in the man-hour tables.

2.2.2 Support types whose fabrication and assembly times are NOT included in Table 9.1

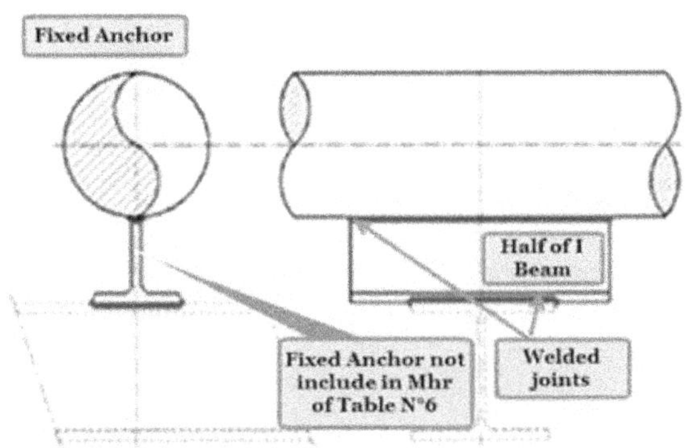

Figure 2.2. Example of a fixed point support NOT INCLUDED in the tables.

2.3 Support Calculation Method

This section describes the method used to estimate the man-hours required for fabricating and installing supports attached to the pipe. This field-verified procedure uses an **equivalent theoretical pipe** model to standardize the estimation and provides realistic results.

This method is only valid for support materials that are in direct contact with the pipe. Table 9.1 of this manual lists the man-hours required for prefabrication and installation of supports per meter (or foot) of pipe, depending on its diameter and wall thickness

The Three-Step Process

To use Table 9.1, first convert the actual pipe run into an **equivalent theoretical pipe**. The process is as follows:

- ➔ **Step 1:** Calculate the total weight of the pipe run. Sum the weights of all components in the run (pipes, fittings, flanges, etc.).
- ➔ **Step 2:** Determine the length of the equivalent theoretical pipe. Divide the total weight (calculated in Step 1) by the unit weight (e.g., kg/m or lb/ft) of the **representative pipe** for that run.
- ➔ **Step 3:** Calculate the total man-hours. Using the diameter and length of the theoretical pipe, refer to Table 9.1 in Chapter 9 of this book to find the man-hour-per-meter (Mhr/m) or man-hour-per-foot (Mhr/ft) ratio. Multiply this ratio by the equivalent length to get the total hours.

2.4 Practical Application Example

Next, the three-step process is applied to calculate the supports for a header from NPS 6" down to the NPS 3" Slip-On flange. The following figure shows this process:

Figure 2.3. This is the piping section for the example.

2.4.1 Step 1: Calculating the Total Weight of the Piping Section

In this step, we calculate the total weight of the pump suction pipe section that runs from the 6" NPS manifold to the 3" NPS slip-on flange (from the previous figure). The results are shown in Table 2.1.

Table 2.1. Calculating Materials and Weights for the Suction Section

Material	Quantity	Unit Weight (kg)	Total Weight (kg)
4" x 3" Eccentric reducer	1.00	1.59	1.59
6" x 4" Reducing Tee	1.00	15.00	15.00
4" NPS 150# SO Flange	2.00	5.44	10.88
3" NPS 150# SO Flange	1.00	4.08	4.08
4" NPS 150# Gate Valve	1.00	64.00	64.00
4" NPS 150# Strainer	1.00	40.00	40.00
4" NPS 90° LR Elbow	1.00	3.86	3.86
4" NPS SCH 40 Pipe	1.50	16.06	24.09
		Grand Total	**163.50**

2.4.2 Step 2: Determining the Equivalent Theoretical Pipe Length

The representative pipe for the suction run is NPS 4" SCH 40, with has a unit weight of 16.06 kg/m.

Divide the total weight of the run by the unit weight of the representative pipe:

$$\frac{163.50\,\text{kg}}{16.06\,\frac{\text{kg}}{\text{m}}} = 10.18\,\text{m}$$

In this case, the equivalent length is 10.18 meters of a theoretical NPS 4" SCH 40 pipe.

2.4.3 Step 3: Calculating the Necessary Man-Hours

Below is a snippet of the Table, which is shown in its entirety in Chapter 9 of this book:

Table 9.1: Table for Piping Support Calculation

NPS	Pipe Weight in kg/m		Support Spacing in meters	Support Weight in kg/m		Support Prefabrication (Mhr/m)		Support Installation (Mhr/m)	
	SCH 40	SCH 80		for SCH 40 pipe	for SCH 80 pipe	for SCH 40 pipes	for SCH 80 pipes	on SCH 40 pipes	on SCH 80 pipes
1"	2.50	3.25	2.14	0.50	0.65	0.50	0.65	0.10	0.13
1 1/2"	4.10	5.40	2.75	0.80	1.00	0.60	0.80	0.10	0.15
2"	5.40	7.47	3.05	1.00	1.35	0.70	0.90	0.20	0.22
2 1/2"	8.50	11.50	3.35	1.30	1.70	0.75	1.00	0.20	0.26
3"	11.30	15.25	5.65	1.50	2.00	0.80	1.10	0.20	0.30
4"	16.06	22.29	4.30	2.00	2.70	0.90	1.20	0.27	0.36
6"	28.30	42.51	5.20	3.00	4.70	1.20	1.80	0.40	0.60

Figure 2.4. Excerpt from the man-hour reference table.

According to the table, the values are as follows:

- ➔ Prefabrication: 0.90 Mhr/m
- ➔ Installation: 0.27 Mhr/m

Add both values to obtain the total rate.

The total is 0.9 + 0.27 = 1.17 hours per meter of pipe.

Multiply the total rate by the equivalent length to find the total man-hour consumption for fabricating and installing the supports for this run.

Total Consumption = 1.17 Mhr/m × 10.18 m = 11.91 Man-Hours.

2.4.4 Analyzing the Result

- ➔ **Scope:** The calculated value (11.91 man-hours) applies exclusively to supports that are directly attached to the pipe, such as lugs, trunnions, pipe shoes, and wear plates.
- ➔ **Exclusions:** This calculation does not include corrosion protection for the supports or man-hours for productivity adjustments.

2.4.5 Productivity

The data in the Table corresponds to standard productivity under the conditions defined in Stage 1 of the previous chapter.

However, each project is unique, and external factors can influence actual labor productivity.

2.5 Calculating Mhr for the Anti-Corrosion Coating of Supports

To estimate the man-hours required for the anticorrosive treatment of supports, it is first necessary to calculate their total surface area. Unlike pipes, this area is determined indirectly from the total weight of the supports.

2.5.1 Calculating the Total Weight of Supports

The calculation starts with two pieces of data: the **equivalent length** of the system (10.18 m) and the **weight of supports per meter** for the reference pipe (NPS 4" SCH 40), which, according to Table 9.1, is 2 kg/m.

Multiplying both values yields the total weight:

$$10.18\,\text{m} \times 2\,\frac{\text{kg}}{\text{m}} = 20.36\,\text{kg} \approx 0.0204\,\text{tons}$$

2.5.2 Calculating the Surface Area to Be Painted

Finally, this weight is converted to a surface area using the **area-to-weight ratio** for lightweight structures (35 m^2/ton).

$$0.0204\,\text{ton} \times 35\,\frac{\text{m}^2}{\text{ton}} = \mathbf{0.714}\,\mathbf{m^2}$$

2.5.3 Calculating Man-Hours

Manual preparation and painting methods were chosen for the small area of 0.714 m^2, with a total of 2.64 man-hours. These methods were power tool cleaning (1.50 m^2/m^2) and traditional brush painting (2.20 m^2/m^2).

These methods are more efficient on a small scale, primarily due to the minimal preparation time required and the high transfer efficiency. Brush application virtually eliminates material waste due to overspray, for example.

For this reason, abrasive blasting and airless spraying were ruled out. Using heavy equipment such as compressors and shot blasters for such a small area incurs disproportionate logistical and installation costs. Furthermore, the need for ample containment space to protect adjacent surfaces renders these industrial methods impractical and uneconomical for specific small-scale projects.Productivity rates for larger batches, where industrial methods become more appropriate and economically viable, are discussed in detail in Chapter 14.

2.6 Estimating the Area for Anti-Corrosion Coating on Piping

Calculating the exact surface area of each fitting (e.g., elbows, tees, and reducers) is a complex and impractical process. Therefore, a method of equivalent calculation length is used, which simplifies the process without sacrificing the necessary accuracy.

This approach is widely accepted in the industry and significantly speeds up the work by converting the geometric complexity of fittings into additional pipe length. This standardization is crucial for ensuring consistency in material procurement and cost estimation, especially on large-scale projects where thousands of such components are used. The conversion factors are based on established engineering tables, providing a reliable balance between precision and on-site practicality.

Important Exclusions. This calculation does not include components such as valves, strainers, or meters. This equipment is usually supplied by the manufacturer with the final paint system already applied, often as part of a specialized, factory-controlled coating process designed to meet specific performance and warranty requirements. Attempting to coat these complex items in the field would be inefficient and could compromise their operational integrity and protective finish.

2.6.1 Area Calculation Method

This pragmatic method streamlines what would otherwise be a highly complex task, ensuring consistency and efficiency in project estimations. This simplified approach is crucial for accurate bidding and material planning, especially on large-scale industrial projects. The process is divided into two clear stages:

Step 1: Determine the Calculation Length. Start with the actual pipe length and add a length (L_e) indicated in:

Chapter 10, Table 10.3: Equivalent Lengths of Fittings for Estimating Painting and Cleaning

For each fitting (elbows, tees, etc.) that needs to be painted, this "calculation length" represents the entire section as if it were a single straight pipe.

$$\text{Calculation Length} = \text{Actual Length} + (\text{Number of Fittings} \times L_e)$$

Stage 2: Calculate the Total Area to be Painted. Next, multiply the total Calculation Length by the pipe's specific external surface area per meter (m^2/m). This **Unit Surface Area** value is sourced from standard engineering handbooks and is based on the pipe's nominal diameter. This final step provides a dependable total area for coating procurement, labor budgeting, and overall project management, forming the basis for cost and schedule control.

$$A_{\text{paint}} = L_{\text{calc}} \times \text{Unit Surface Area}$$

2.6.2 Applying Productivity Rates

Once the total area is determined, the next step is to multiply it by the man-hour per square meter coefficient (Mhr/m^2). Productivity rates vary depending on the scale of the work; rates for medium-sized batches are detailed in Table 10.2 (Chapter 10), while small batches require specific analysis, as seen in 2.5.3.

Chapter 3

Calculating Man-Hours for Piping

The calculation of standard man-hours required to install a complete piping line consists of three fundamental elements:

- ➔ **Joints:** All joints in the system (welded, flanged, or threaded) are quantified and multiplied by their respective Mhr value from the reference tables.
- ➔ **Pipe Handling and Positioning:** This corresponds to the hours required for handling, presenting, and installing the straight sections of pipe.
- ➔ **Component Installation:** This includes the individual installation of valves, strainers, instruments, and any other accessory that is part of the line.

The sum of these three groups provides our starting point for the estimation.

3.1 From Standard Hours to a Realistic Estimate: Adjustment Factors

All man-hour values presented in the tables of this manual represent the effort required to execute a project under ideal conditions. We will refer to this base value as **Tabulated Standard Hours (TSH)**.

The validity of the TSH depends on the project conditions being comparable to the standard ones,which assume the following:

- ➔ Personnel perform their tasks at a normal and efficient pace.
- ➔ Standard tools and equipment are used.
- ➔ The worksite is isolated from interference from other operating plants.
- ➔ The materials are identical to those specified in the tables.

However, in practice, no project is perfectly **standard**. The actual conditions of the worksite, logistics, weather, or the use of advanced technology all influence labor productivity.

Therefore, **the TSH is our starting point, not the final result**. This base value must be adjusted using **adjustment factors** that reflect the unique characteristics of each project. These factors quantify the expected deviation from standard productivity and can either increase or decrease the final number of estimated hours.

3.2 Types of Corrections

The total Tabulated Standard Hours (TSH) are subject to at least four types of adjustments that modify the final value:

- ➔ **Project Variables Correction:** This adjustment is based on specific management, environmental, and resource conditions (analyzed in detail below).
- ➔ **Site Condition Adjustment:** This adjustment is based on the geographical location and logistics of the worksite.
- ➔ **Material Type or WPS Condition Correction:** A coefficient (k) for adjustments based on material or other conditions defined by the Welding Procedure Specification (WPS).
- ➔ **Prefabrication Correction:** An adjustment based on the degree of pipe prefabrication in a workshop.

3.3 Analysis of Project Variables

Below are some of the most common variables that affect productivity. After evaluating the answers to the key questions, the impact of each variable is weighed and assigned a performance rating: Very Poor, Poor, Standard, or Excellent.

3.3.1 Contractor Experience

This variable measures the experience level of the bidder and their team in executing and supervising similar projects.

Key Questions:

- ➔ Has the company executed projects of this type and size? How frequently?
- ➔ What is the experience of the proposed supervisory team for this type of work?

- ➔ Does the company have access to qualified field personnel with proven experience?

3.3.2 Quality of Bidding Information

This measures the accuracy of the documentation available for preparing the estimate.

Key Questions:

- ➔ Is the quality of the preliminary engineering sufficient to bid with an acceptable level of risk?
- ➔ Was a site visit conducted with experienced personnel?
- ➔ Were all aspects that could influence productivity (access, interferences, utilities, etc.) reviewed and documented during the visit?

3.3.3 Labor and Union Relations

Analyzes the complexity and constraints of the regulatory and union relations governing the project area and its potential impact on project execution.

Key Questions:

- ➔ What will the work schedule be? Will it include night shifts, when productivity is typically lower?
- ➔ What is the local union's history of behavior on previous projects?
- ➔ What hiring requirements does the collective bargaining agreement impose (such as the obligation to use local hiring halls), and how could this influence productivity?

3.3.4 Site Conditions and Access

Evaluate the physical characteristics of the terrain and the ease of access for personnel, materials, and equipment.

Key Questions:

- ➔ What are the ground conditions at the site? Is it low-lying and prone to flooding, or high and dry?
- ➔ Will the distance and condition of access roads hinder personnel commutes and material logistics?
- ➔ Are there access or circulation restrictions within the worksite itself?

3.3.5 Climatic Conditions

Analyze the impact of the typical climate in the project area on the productivity of outdoor work.

Key Questions:

- ➔ What is the historical weather pattern for the area during the project execution period?
- ➔ Does it coincide with seasons of high probability of rain, wind, snow, or extreme temperatures?
- ➔ Could the local climate, if significantly different from what personnel are accustomed to, affect their acclimatization and performance?

3.3.6 Working at Height

Important Context: The man-hour values in the tables correspond to tasks performed on working planes no higher than 1.50 meters (approximately 5 feet) above the ground. Any work above this elevation requires an adjustment. **Key Questions:**

- ➔ What percentage of the work will be executed above the standard height of 1.50 m?
- ➔ Will the extensive use of scaffolding significantly slow down personnel movement, material handling, and equipment access to the work point?
- ➔ Should additional time or special logistics be considered that are not included in the standard installation hours?

3.4 Other Key Correction Factors

Geographic Location Correction

- ➔ **Critical logistics:** Delivery delays and transit damage risks for specialized materials (valves, welding consumables), coupled with dependency on complex freight routes.
- ➔ **Local craft talent deficit:** Necessity to import specialized labor (e.g., ASME Section IX certified 6G welders), driving up turnover, absenteeism, and extended field learning curves.
- ➔ **Deficient infrastructure:** Continuous reliance on temporary diesel generator sets, telecommunication dropouts, and limited on-site emergency medical response.
- ➔ **Camp & rotation fatigue:** Significant productivity degradation toward the end of shift rotations $(14 \times 7, 21 \times 7)$ due to accumulated physical fatigue and upcoming leave anticipation.
- ➔ **Hostile environment:** Extreme physiological and climatic conditions (high-altitude hypoxia, extreme heat, or tropical humidity) that impair craft stamina, equipment efficiency, and material integrity.

Site Condition Correction (Brownfield vs. Greenfield)

This multiplier distinguishes between new construction executed on unobstructed open terrain (*greenfield*) and revamp modifications performed within the active battery limits of an operating industrial plant (*brownfield*).

Working within a *brownfield* environment drastically reduces real craft productivity, as a significant portion of the shift is absorbed by Non-Productive Time (NPT), including:

- ➔ **Permit-to-Work (PTW) management:** Standby waiting time for daily work authorization, mandatory atmospheric gas testing, and shift coordination with plant operations.
- ➔ **Safety and logistical restrictions:** Strict adherence to facility safety protocols (*LOTO*, permit-required confined spaces, pressurized hot-work habitats), imposing an inherently slower pace and severely congested logistics.
- ➔ **Technical complexity:** High-risk execution of critical tie-ins and final tie-in welds onto live process piping systems.

All the above factors introduce operational risk, continuous safety oversight, and workspace constraints that simply do not exist in an isolated *greenfield* installation.

Author's Field Experience

During piping revamp projects executed inside operating refineries (within classified fire-hazard and explosive zones), direct man-hour consumption exceeded standard baseline catalogue values by 25%.

In contrast, for construction packages executed for the same refinery but located on unobstructed greenfield plots, the variance was only 3% to 5% over the standard baseline.

3.4.1 Correction for Material Type and WPS

The TSH values apply to standard carbon steels (e.g., ASTM A53). If other materials (alloy steels, stainless steels) are used, or if the Welding Procedure Specification (WPS) defines more demanding parameters, the TSH must be adjusted with a conversion factor (k).

3.4.2 Correction for Prefabrication Strategy

In industrial projects like refineries or chemical plants, piping installation is one of the most critical disciplines.

The most effective strategy to reduce costs, ensure quality, and meet deadlines is the prefabrication of piping.

Fundamental Principle: Anything fabricated in a workshop is more economical and of higher quality than on-site work.

The success of a project largely depends on the quantity and quality of prefabrication.

This strategy divides the work into two phases: first, the prefabrication of pipe segments (spools) in a workshop; then, their transport, installation, and final fit-up in the field.

Key to Success: Timely delivery of pipe spools to the installation site is crucial for project success.

3.4.3 The Prefabrication Process

Spools are fabricated in specialized workshops or temporary site facilities from pipes and fittings (elbows, flanges, etc.).

Hiring suitable vendors for this task ensures reduced costs and a higher-quality final product.

These workshops have the necessary infrastructure, quality management systems, and personnel to guarantee dimensional accuracy and strict compliance with technical specifications

3.4.4 Key Advantages of Prefabrication

Prefabrication generates significant savings in labor, inspection, and testing. Its main advantages are:

- ➔ Improved Quality Control: It is easier to maintain high standards in a controlled workshop environment.
- ➔ Dimensional Accuracy: Adherence to strict tolerances in the workshop prevents costly corrections in the field.
- ➔ Weather Independence: Fabrication continues without interruption from rain, wind, or extreme temperatures, which eliminates delays.
- ➔ Labor Optimization: It addresses potential shortages of skilled labor at the project site and reduces the need to mobilize large fabrication teams in the field.
- ➔ Economies of Scale: Mass production in an optimized environment reduces fabrication costs compared to on-site work.
- ➔ Schedule Reduction: Faster field installation translates to lower indirect costs and meeting deadlines.

- ➔ Lower Investment: The installing contractor does not need to invest in expensive fabrication and testing equipment.
- ➔ Nondestructive Examination (NDE): Tests such as radiography, magnetic particle testing, and liquid penetrant testing are performed more efficiently and safely in a workshop.
- ➔ Higher Weld Quality: Control over welding parameters is superior in a workshop, resulting in a lower repair rate.
- ➔ Safer and Cleaner Work Environment: It reduces personnel exposure to hazards and minimizes contamination at the job site, a critical factor in food or pharmaceutical plants.

3.4.5 Precautions and Associated Risks

Despite its enormous benefits, prefabrication presents a severe risk: dimensional errors.

A poorly fabricated spool can trigger a chain of corrections throughout the entire line, forcing modifications to other segments and repetition of costly tests and inspections.

The most frequent causes are misinterpreting isometrics in the workshop or errors in detailed engineering.

Impact on Productivity: When a prefabrication strategy is adopted, man-hour consumption for tasks such as welding joints is significantly reduced.

This reduction depends on factors such as the workshop's level of automation, and the pipe diameter and thickness.

3.5 Case Study: Comparative Analysis by Geographic Region

For example, we analyze the same bid for projects planned in three different geographic regions (A, B, and C), each with its own unique characteristics.

The objective is to demonstrate how the estimating team applies a variable analysis methodology to quantify the impact of local conditions. Starting from a base of standard hours, the team determines adjustment factors to reflect the risks and opportunities of each location. This achieves a much more accurate and defensible estimate. This analysis is essential for grasping why labor costs for the same job can differ significantly depending on the location.

3.5.1 Qualitative Analysis of Variables

For each region, six critical variables affecting productivity are evaluated. Each variable is rated against a standard condition. The following table summarizes the qualitative assessment.

Table 3.1. Comparative Summary of Variables by Region

Variable	Region A	Region B	Region C
1. Contractor Experience	Excellent (1.15)	Excellent (1.15)	Excellent (1.15)
2. Quality of Documentation	Standard (1)	Standard (1)	Standard (1)
3. Labor & Union Environment	Standard (1)	Adverse (moderate variations) (0.8)	Very Adverse (significant variations) (0.6)
4. Site Conditions	Standard (1)	Poor (poorly drained soils) (0.8)	Standard (1)
5. Climatic Conditions	Adverse (0.8)	Adverse (0.8)	Very Adverse (constant winds >18.6 mph) (0.7)
6. Work at Height	Adverse (working elevation at 8.2 ft) (0.7)	Standard (1)	Standard (1)

3.5.2 Quantitative Summary and Conclusion

The quantitative analysis culminates in the calculation of an average productivity factor for each region. This overall factor, obtained from the arithmetic mean of the six variables, and is used to adjust the base estimate (TSH) to reflect expected performance at each location. Note that an average factor **below 1.00** indicates lower-than-standard productivity, which translates into an **increase** in the required man-hours.

The results for each region are:

- **Region A:** With an average factor of **0.94**, a lower-than-ideal performance is projected, requiring a **6%** increase in the estimated man-hours.
- **Region B:** Its average factor of **0.925** reflects a more significant impact from local conditions, resulting in an upward adjustment of **8.1%** to the base man-hours.

➔ **Region C:** Being the most unfavorable option, its average factor of **0.91** implies the greatest impact on productivity, requiring a **10%** increase in the hour estimate to reflect the actual site conditions.

Strategic Implications of the Analysis. These results demonstrate that although the technical scope of work is identical, the actual cost and risk profile of the project varies dramatically by geography. A difference of nearly 4 percentage points in man-hours between Regions A and C could determine whether a company wins or loses a competitive bid.

Beyond direct costs, this analysis functions as a risk management tool. It reveals that climatic and union factors are the main sources of uncertainty that could cause additional delays in Region C. Therefore, the final bid must reflect not only a higher number of hours, but also an appropriate execution strategy and contingency plan to mitigate the specific risks of the least favorable region.

Chapter 4

Assembling Above-Ground Piping

4.1 Objective and Scope

The objective of this chapter is to provide man-hour consumption rates per linear meter for the assembly of piping in industrial plants, specifically for above-ground installations.

Table 4.1 of this manual lists the man-hours to be used according to the nominal pipe size (NPS) and the schedule (SCH) of the pipe.

The values in this table apply to standard-grade carbon steel pipes, such as:

➔ ASTM A-53 / Grade A / B.

➔ ERW A-53 / API 5L PSL2 Grade B.

Note on other materials: For steels other than those specified (e.g., A106, A333, low alloy steels, stainless steels, or high yield strength steels), or for welding conditions specified in the WPS, the times in Table 4.1 shall be adjusted by multiplying them by the correction factor (k).

4.2 Fundamental Definitions

Piping (Piping System) Is defined as a complex network of pipes, fittings, and components, designed according to standards such as ASME B31.3 (process plants) or ASME B31.1 (power plants), used to transfer fluids between different pieces of equipment within the boundaries of a plant.

NPS (Nominal Pipe Size) It is the North American standard designation for the size of a pipe, expressed in inches. It is important to note that the NPS value does not always coincide with the actual outside diameter.

- ➔ **For NPS 12 and smaller:** The actual outside diameter is larger than the nominal value.
- ➔ **For NPS 14 and larger:** The actual outside diameter coincides with the nominal value.

ASTM Grade The standards of the American Society for Testing and Materials (ASTM) define the chemical composition and the manufacturing process. The **Grade** (e.g., Grade A, Grade B) specifies the allowable percentages of carbon, **manganese**, nickel, and other elements, thus determining the mechanical properties of the material.

Pipe Lengths Pipes are supplied in standard commercial lengths. For calculation purposes, it is advisable to use the following measurements, as other sizes may have a higher cost:

- ➔ **Single Random Length:** Between 5 and 7 meters.
- ➔ **Double Random Length:** Between 11 and 13 meters.

Types of Pipe Ends There are three standard versions for the ends, each associated with a type of joint:

- ➔ **Plain Ends:** For *Socket Weld* or *Slip-On* fittings.
- ➔ **Threaded Ends:** For threaded fittings and flanges in small diameters.
- ➔ **Beveled Ends:** Standard for butt-welded joints. The bevel angle is 30° ($+5^\circ$/-0°) with a root face of 1.6 mm (±0.8 mm).

4.3 Fabrication and Tolerances

Figure 4.1 indicates the dimensional tolerances allowed in Piping fabrication.

1. Variations in the dimensions indicated in the drawing, between flange faces and centers of fittings, cannot exceed 3 mm (the tolerance is cumulative).
2. In bends with a radius $\geq 6\phi$, variations of the surface (wrinkles, bulges) shall not exceed 3% of the nominal pipe diameter.
3. The lateral translation of the flanges from the position indicated in the drawing must not exceed 1.5 mm.
4. The rotation of the flanges must not exceed 1.6 mm on the perimeter of the bolt hole.

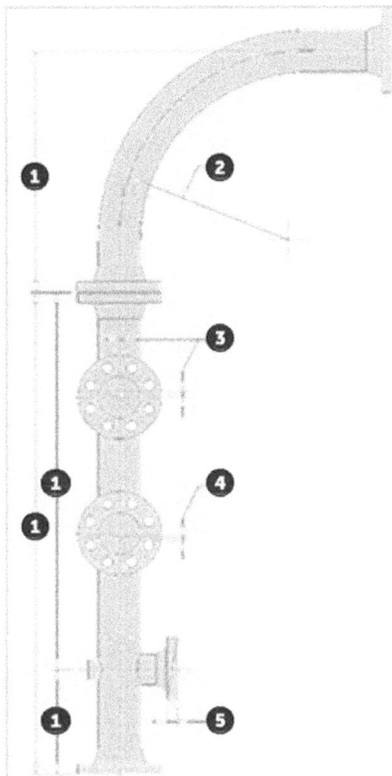

Figure 4.1. Fabrication and Assembly Tolerances for Piping.

5. The alignment of the flange faces must not deviate more than 0.80 mm.

4.4 Man-Hours for Assembling Above-Ground Piping

This section details the use of the data from **Table 4.1**, which considers the assembly of straight sections of commercial length.

4.5 Man-Hour Data and Application

This section provides man-hour consumption ratios for above-ground piping installation, including an explanation of how to use the provided data table.

Scope of the Table Values. It is essential to understand which tasks are included and excluded in the man-hour rates.

Included Tasks Complete handling (unloading, storage, transport to the assembly point), alignment on existing supports, joint welding, and leak testing.

Excluded Tasks Fabrication and installation of supports, cleaning, painting, or application of thermal insulation.

References and Applicable Standards. The values are consistent with the dimensions and tolerances of the **ASME B36.10**, **ASME B36.19**, **ASTM A530/A530M**, and **ASTM A999/A999M** standards.

Table 4.1. Mhr for Above-Ground Piping Installation (Handling, Alignment, Welding, and Testing)

	SCH 10		SCH 20		SCH 30		SCH 40		SCH 60		SCH 80		SCH 160	
NPS	kg/m	Mhr	kg/m	Mhr	kg/m	Mhr	kg/m	Mhr	kg/m	Mhr	kg/m	Mhr	kg/m	Mhr
1/2"	–	–	–	–	–	–	1.30	–	–	–	1.60	–	1.90	–
3/4"	–	–	–	–	–	–	1.70	–	–	–	2.20	–	2.90	–
1"	–	–	–	–	–	–	2.50	–	–	–	3.20	–	4.20	–
1 1/2"	–	–	–	–	–	–	4.10	–	–	–	5.40	–	7.20	–
2"	–	–	–	–	–	–	5.50	0.60	–	–	7.50	0.70	11.10	0.80
2 1/2"	–	–	–	–	–	–	8.50	0.70	–	–	11.50	0.90	14.90	1.00
3"	–	–	–	–	–	–	11.30	0.90	–	–	15.30	1.10	21.30	1.30
4"	–	–	–	–	–	–	16.06	1.00	–	–	22.30	1.30	33.50	1.60
6"	–	–	–	–	–	–	28.60	1.50	–	–	43.00	1.90	67.40	2.70
8"	–	–	33.00	1.70	37.00	1.80	43.00	2.00	53.00	2.30	65.00	2.50	114.00	4.10
10"	–	–	42.00	2.00	51.00	2.20	60.00	2.50	82.00	3.00	96.00	3.50	172.00	6.00
12"	–	–	20.00	2.30	65.00	2.70	80.00	3.00	109.00	3.80	132.00	4.40	239.00	7.90
14"	55.20	2.30	68.50	2.70	81.80	3.20	93.70	3.60	126.00	4.50	160.00	5.40	283.00	–
16"	62.60	2.50	78.00	3.00	93.10	3.60	123.00	4.30	160.00	5.40	204.00	6.90	365.00	–
18"	70.50	2.80	88.00	3.50	122.00	4.20	156.00	5.20	206.00	7.00	254.00	8.40	460.00	–
20"	78.50	3.00	117.00	4.10	155.00	5.20	183.00	6.20	248.00	8.00	311.00	10.00	564.00	–
24"	94.00	3.50	141.00	5.00	210.00	6.60	255.00	8.00	354.00	11.00	441.00	14.00	806.00	–

4.6 Adjustment Factors for Alternative Scenarios

For scenarios other than the initial assembly, apply the following adjustment factors to the values in Table 4.1:

➔ **For Disassembly or Removal:** Multiply the values by a factor of **0.50**.

→ **For Relocation (Disassembly and Assembly):** Multiply the values by a factor of **2.0**. (Note: this factor covers the complete cycle of dismantling the pipe and reinstalling it in a new location).

Chapter 5

Installing Elbows and Tees

5.1 Objective and Scope

- The objective of this chapter is to determine the number of man-hours required for the field assembly of long-radius elbows (90° and 45°) and Straight Tees in above-ground piping installations.
- This data is found in Table 5.1 of this chapter, where the times are classified according to the nominal pipe size (NPS) and the schedule (SCH) of the fittings.
- The values presented apply to standard-grade carbon steel fittings manufactured in accordance with the ASTM A234 Gr. WPB standard.

Note on other materials: For steels other than those specified (e.g., A106, A333, low-alloy steels, stainless steels, or high-yield steels), or for specific welding conditions established in the WPS, the times in the Table must be adjusted by multiplying them by the correction factor (k).

5.2 Definitions and Technical Notes

- **Fitting Wall Thickness.** Generally, forged fittings have a slightly greater wall thickness than the corresponding pipe. This is to compensate for any potential thickness reduction that may occur during the forming process, thereby ensuring the structural integrity of the component.
- **Fitting Material.** The material of the fittings must always be compatible with, and generally match, that of the pipes to which they will be joined.

Note on Wall Thickness Nomenclature: It is important to understand the relationship between the SCH, STD (Standard), and XS (Extra Strong) designations:

➔ STD vs. SCH 40: They are equivalent up to NPS 10. For larger diameters, STD corresponds to a fixed wall thickness of 3/8" (9.53 mm).

➔ XS vs. SCH 80: They are equivalent up to NPS 8. For larger diameters (NPS 10 and above), XS corresponds to a fixed wall thickness of 1/2" (12.7 mm).

Table 5.1. Installation of LR Elbows and Normal Tees (Man-Hours)

	LR 90° AND 45° ELBOWS				**NORMAL TEE**		
NPS	**Light**	**STD**	**Ex. Heavy**	**SCH 160**	**STD**	**Ex. Heavy**	**SCH 160**
1/2"	–	1.40	1.80	2.00	2.10	2.70	3.00
3/4"	1.60	1.60	2.00	2.20	2.40	3.00	3.30
1"	1.80	1.80	2.20	2.80	2.70	3.50	3.90
1 1/2"	2.40	2.40	2.80	3.30	3.60	4.20	5.00
2"	2.80	2.90	3.40	4.00	4.40	5.10	6.00
2 1/2"	3.10	3.50	4.00	5.00	5.10	6.00	7.50
3"	3.60	4.00	5.00	6.00	6.00	7.70	9.00
4"	4.20	5.00	6.20	8.00	7.50	9.30	12.00
6"	5.80	7.00	9.00	13.00	11.00	14.00	20.00
8"	7.50	9.50	13.00	21.00	14.00	19.00	32.00
10"	9.60	12.00	15.00	32.00	18.00	22.00	48.00
12"	12.00	14.00	17.00	45.00	20.00	26.00	66.00
14"	–	15.00	19.00	–	22.00	28.00	–
16"	–	17.00	21.00	–	24.00	31.00	–
18"	–	18.00	23.00	–	27.00	36.00	–
20"	–	20.00	26.00	–	30.00	39.00	–
24"	–	23.00	32.00	–	34.00	48.00	–

The Table presents the estimated man-hours required for the assembly of 90° and 45° long-radius (LR) elbows, as well as Straight Tees, in installations located above ground level.

The values vary according to the nominal diameter and Schedule (SCH) of the fittings, and are applicable to carbon steel parts conforming to the ASTM A234 Gr. WPB standard.

5.3 Scope of Table Values

Included Tasks:

➔ Handling: Unloading the fitting on-site, storage, and subsequent transport to the installation point.

➔ Fit-up and Alignment: Positioning the fitting in the line.

➔ Welding of the ends to the pipe.

➔ Leak testing of the completed joints.

Explicitly Excluded Items:

➔ Fabrication and installation of supports.

➔ Cleaning, painting, and thermal insulation work.

➔ Prefabrication of fittings (it is assumed that the fittings are standard catalog components).

5.4 Welding Requirements for Olet-Type Branch Connections

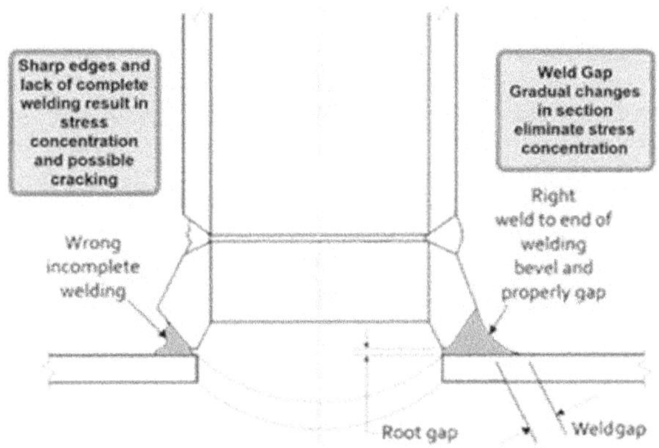

Figure 5.1. Quality control of welding in Olet-type branches.

The illustration above details the correct and incorrect weld profiles for the installation of Olet-type branch fittings (e.g., Weldolet®, Sockolet®).

Applicable design codes, such as ASME B31.3 (Process Piping) and the ASME Pressure Vessel Code, establish the specific requirements for structural reinforcement and welding of these connections.

It is critically important to follow the proper procedure, as poor quality or an incorrect profile can create high-stress concentrations and lead to fatigue failure or cracking at the root of the bead, which compromises the mechanical integrity of the entire connection.

Chapter 6

Installing Reducing Tees, Saddles, and Reducers up to 10" $\times$ 4"

6.1 Objective and Scope

The objective of this chapter is to calculate the number of man-hours required for the assembly of the following fittings in above-ground installations:

➔ Reducing Tees

➔ Concentric and Eccentric Reducers

➔ Saddles

The values in Table 6.1 apply to standard-grade ASTM A234 Gr. WPB carbon steel fittings.

Note on Material Adjustments:

For steels other than the specified baseline material (e.g., A106, A333, low-alloy, stainless, or high-yield steels), or for unique conditions defined in the Welding Procedure Specification (WPS), the times provided in the Table must be adjusted. This is accomplished by multiplying the baseline time by the material correction factor (k).

This factor is essential as it compensates for critical differences in the inherent properties of various alloys, primarily their thermal conductivity and hardenability. For instance, high-alloy materials frequently require substantially longer post-weld heat treatment (PWHT) soak times to ensure proper tempering and stress relief. To ensure procedural accuracy, always consult the governing code and the project-specific WPS to determine the precise k-value required for the application.

latex

Table 6.1. Mhr for Installation of Reducing Tees, Reducers, and Saddles up to 10 x 4 inches

	Reducing Tee			Concentric and Eccentric Reducers			Saddles
NPS	Sch 40 Mhr	Ex. Heavy Mhr	Sch 160 Mhr	Sch 40 Mhr	Ex. Heavy Mhr	D. Ex. Heavy Mhr	Mhr
1" x 3/4"	2.60	3.20	4.40	1.70	2.10	2.40	–
1 1/4" x 1"	2.80	3.50	4.70	1.90	2.30	2.60	–
1 1/2" x 1 1/4'"	3.10	3.90	5.00	2.10	2.50	2.90	–
2" x 1 1/2"	4.00	4.80	5.70	2.70	3.10	3.70	–
2 1/2" x 2"	4.90	5.70	7.00	3.20	3.70	4.50	–
3" x 1 1/2"	5.20	6.40	7.70	3.30	4.00	4.70	–
3" x 2"	5.40	6.70	8.00	3.50	4.20	5.00	5.00
3" x 2 1/2"	5.80	7.00	8.60	3.70	4.50	5.50	5.50
4" x 2"	6.50	7.90	10.00	4.00	4.80	6.00	6.00
4" x 2 1/2"	6.70	8.20	10.50	4.20	5.10	6.50	6.50
4" x 3"	7.00	8.70	11.00	4.50	5.60	7.00	7.00
6" x 3"	9.00	11.50	16.00	5.50	7.00	9.50	11.00
6" x 4"	9.50	12.10	17.00	6.00	7.60	10.50	12.00
8" x 4"	12.00	15.50	25.00	7.20	9.30	14.50	15.00
8" x 6"	13.00	17.00	28.00	8.20	10.70	17.00	17.00
10" x 4"	14.00	18.00	36.00	8.50	11.00	20.50	18.00

6.2 Scope of Table Values

Included Tasks:

➔ Handling: Unloading the fitting on-site, storage, and subsequent transport to the installation point.

➔ Fit-up and Alignment: Positioning the fitting in the line.

➔ Welding of the ends to the pipe.

➔ Leak testing of the completed joints.

Explicitly Excluded Items:

➔ Fabrication and installation of supports.

➔ Cleaning and painting work.

➔ Application of thermal insulation.

6.3 General Guidelines

Reducing Tees: Availability and Alternatives

While Straight Tees are available in virtually all diameters, the same is not true for Reducing Tees.

Many sizes are not standardly manufactured due to production complexity or low demand.

For example, an NPS 6"× 4" tee is a common item, but it is very unlikely to find an NPS 16"× 2" tee.

6.3.1 Fabrication Alternatives Accepted by ASME B31.3

Even if it existed, using an NPS 16 inch tee for an NPS 2 branch would be neither technically nor economically practical. In these cases, the appropriate solution is to use a branch connection, which can be achieved by:

➔ **Stub-in Branch Connection:** This involves cutting a hole in the main pipe and then inserting and welding the branch.

  ➢ *Limitation:* This method weakens the header, which almost always requires a reinforcement calculation according to ASME B31.3 (paragraph 304.3.3).

  ➢ *Considerations (if reinforcement is required):* If the calculation demands the addition of a reinforcement pad (saddle), it must be made of a compatible material and include a weep hole.

➔ **Branch Fitting (Olet):** This is the most robust solution, especially for high-pressure/high-temperature service. Forged fittings such as Weldolet®, Sockolet®, etc., are used. This topic is covered in the next chapter.

Chapter 7

Installing Reducing Tees, Saddles, and Reducers up to 20” $\times$ 18”

7.1 Objective and Scope

Objective The objective of this chapter is to determine the number of man-hours required for the assembly of large-diameter fittings in above-ground installations. The Table covers the following components up to a size of 20 x 18 inches:

- ➔ Reducing tees
- ➔ Concentric and eccentric reducers
- ➔ Saddles

The values in Table 7.1 apply to standard-grade ASTM A234 Gr. WPB carbon steel fittings.

Note on other materials: For steels other than those specified (e.g., A106, A333, low-alloy steels, stainless steels, or high-yield steels), or for specific welding conditions established in the WPS, the times in Table 7.1 must be adjusted. This is done by multiplying them by the correction factor (k).

latex

7.2 Scope of Table Values

Included Tasks:

Table 7.1. Man-Hours for Installation of Large-Diameter Fittings

	Reducing Tee			**Concentric and Eccentric Reducers**			**Saddles**
NPS	Sch 40 (Mhr)	Extra Heavy (Mhr)	Sch 160 (Mhr)	Sch 40 (Mhr)	Extra Heavy (Mhr)	Double Extra Heavy (Mhr)	(Mhr)
10" x 6"	16.00	20.00	39.00	9.50	12.00	23.00	19.00
10" x 8"	16.00	21.00	42.00	10.00	14.00	27.00	20.00
12" x 6"	17.00	22.00	52.00	10.00	13.00	29.00	21.00
12" x 8"	18.00	23.00	56.00	11.00	15.00	33.00	23.00
12" x 10"	19.00	25.00	60.00	13.00	16.00	39.00	24.00
14" x 8"	19.50	25.00	–	13.00	17.00	–	25.00
14" x 10"	21.00	26.00	–	14.00	18.00	–	26.00
14" x 12"	22.00	27.00	–	14.00	19.00	–	27.00
16" x 12"	23.00	29.00	–	15.00	20.00	–	28.00
16" x 14"	24.00	30.00	–	16.00	21.00	–	29.00
18" x 12"	25.00	32.00	–	16.00	21.00	–	31.00
18" x 14"	26.00	33.00	–	17.00	22.00	–	32.00
18" x 16"	27.00	34.00	–	18.00	22.00	–	33.00
20" x 14"	28.00	36.00	–	18.00	22.00	–	35.00
20" x 16"	29.00	37.00	–	19.00	24.00	–	36.00
20" x 18"	30.00	38.00	–	20	25	–	37

- ➔ Handling: Unloading the fitting on-site, storage, and subsequent transport to the installation point.
- ➔ Fit-up and Alignment: Positioning the fitting in the line.
- ➔ Welding of the ends to the pipe.
- ➔ Leak testing of the completed joints.

Explicitly Excluded Items:

- ➔ Fabrication and installation of supports.
- ➔ Cleaning and painting work.
- ➔ Application of thermal insulation.

7.3 Branch Connection Fittings (Olets)

These fittings allow for reinforced branches from a main pipe without the need for saddles. They are standardized by MSS SP-97 and are available in multiple materials and sizes.

7.3.1 Main Types of Olets

- ➔ **Weldolet:** The most common type. It has beveled ends and is welded directly onto the header pipe. The branch pipe is then butt-welded to the outlet of the Weldolet. Its body inherently includes all necessary reinforcement.
- ➔ **Sockolet:** Similar to a Weldolet, but the connection with the branch pipe is a socket weld. It is generally used for smaller branch diameters (up to 2 inches).
- ➔ **Thredolet:** The outlet has a threaded connection (NPT). It is used for low-pressure and non-critical services.
- ➔ **Latrolet:** A fitting that creates an angled branch connection, typically 45°, on a straight main pipe. Unlike the Weldolet, which creates a 90° branch, the Latrolet is used when the design requires a non-perpendicular insertion.
- ➔ **Elbolet:** Specifically designed to make a branch connection directly from the body of a pipe elbow, either long or short radius. Its base has a pronounced curvature (saddle shape) that adapts perfectly to the radius of the elbow.
- ➔ **Nipolet:** A branch fitting that integrates an extra-strong (XS) pipe nipple into a single forged piece. It combines the function of a Weldolet and a nipple, eliminating the need for a weld between them. Commonly used for instrumentation connections.
- ➔ **Sweepolet:** Provides a smooth flow transition from the header to the branch, which minimizes turbulence and erosion. Its design is ideal for handling high pressures, temperatures, and cyclic or high-vibration services.

Chapter 8

Installing Valves

8.1 General Installation Conditions

8.1.1 Objective

The objective of this chapter is to establish the fundamentals for estimating the man-hours required for valve installation. The values included in the tables correspond to performance achieved under standard application conditions on an isolated worksite, free from interference from operating plants or equipment.

As previously indicated, these performance rates must be adjusted using correction factors when the actual project conditions differ from the standard, in order to accurately reflect the specific situation of each job site.

8.1.2 Standard Application Conditions

The values in the tables were recorded under the following criteria:

- ➔ **Climate:** Tasks are performed in a moderate climate (average temperatures between 5°C and 30°C) and with winds below 30 km/h.
- ➔ **Work Environment:** The area is free of contamination, and work is carried out during daylight hours.
- ➔ **Working Height:** The maximum height of the work plane does not exceed 1.50 meters above ground level. Any installation at a greater height will require an adjustment.
- ➔ **Clearance:** In roofed areas, the minimum free height is 2.60 meters.

- ➔ **Transportation:** The maximum distance from the storage area to the installation point is 300 meters.
- ➔ **Access:** Operator access is 200 meters or less from the work area.
- ➔ **Resources:** Qualified personnel and appropriate equipment and tools are available.

8.1.3 Scope and Table Exclusions

The man-hour records in the Table do not include the following items:

- ➔ Scaffolding assembly and removal.
- ➔ Rework or corrections due to design errors.
- ➔ Hours for the initial adjustment or re-torquing of bolts on flanged connections.
- ➔ Hours for leak testing of large-sized valves prior to their installation.

8.2 Installing Gate Valves

This table indicates the man-hour times for the installation of gate valves in above-ground facilities. The data is classified according to the valve's diameter (NPS) and series (pressure class).

8.2.1 Important Notes on the Table

8.2.1.1 Valve Weight

The weights indicated in the table are approximate. Once the specific model is selected, the manufacturer's catalog must be consulted to obtain the exact weight, which is crucial for planning lifting maneuvers.

8.2.1.2 Valve Type by Diameter

- ➔ **NPS 1/2" to 2":** The values correspond to forged steel valves with threaded ends.
- ➔ **NPS 2 1/2" and larger:** The values correspond to cast steel valves with flanged ends.

8.2.2 Man-hours for Gate Valve Installation

Table 8.1. Gate Valve Installation

NPS	150#		300#		600#		900#		1500#	
	kg	Mhr	kg	Mhr	kg	Mhr	kg	Mhr	kg	Mhr
1/2"	1.25	1.30	1.80	1.40	2.90	1.50	4.50	1.60	6.50	1.70
3/4"	2.50	1.50	2.90	1.60	4.00	1.70	6.00	1.80	10.80	2.00
1"	3.60	1.80	4.50	1.90	6.30	2.00	10.60	2.20	18.00	2.50
1 1/2"	8.40	2.60	9.80	2.70	15.00	2.70	25.40	3.00	38.50	3.50
2"	13.60	3.00	17.00	3.20	19.50	3.40	36.00	3.70	68.00	4.00
2 1/2"	30.00	1.50	47.00	1.70	77.00	2.20	90.00	2.50	164	3.00
3"	41.00	1.70	44.00	2.00	63.00	2.50	131	3.00	289	4.00
4"	64.00	2.00	71.00	2.50	120	3.00	261	4.00	385	5.30
6"	92.00	3.00	127	4.00	270	6.00	515	8.50	784	11.00
8"	147	4.00	196	5.60	430	8.50	891	12.00	1485	16.00
10"	186	5.20	300	7.50	740	12.00	1042	16.00	2595	21.00
12"	289	6.50	407	9.50	874	15.00	2080	20.00	3580	25.00
14"	485	8.00	782	12.50	1300	19.00	2760	25.00	–	–
16"	636	10.00	1043	15.00	1800	24.00	3158	30.00	–	–
18"	840	12.00	1383	18.00	2008	33.00	3762	40.00	–	–
20"	1015	14.00	1864	22.00	2700	45.00	–	–	–	–
24"	1521	18.00	2270	32.00	4500	55.00	–	–	–	–

8.2.3 Installation Recommendations

- ➔ **Preparation:** Inspect the valve for damage, confirm it meets the service specifications, and clean both the valve and the pipe. Verify that the valve mechanism opens and closes smoothly.
- ➔ **Installation:** Keep the valve partially open during installation to protect its internal components from undue stress. Ensure the piping is correctly aligned and supported.

8.3 Installing Wafer Valves

The following Table details the man-hour times for the installation of wafer-type valves in above-ground facilities.

The data is classified by the valve's diameter (NPS) and series. **Note:** The values do not include time for bolt adjustment or re-torquing.

Table 8.2. Man-hours for Wafer Valve Installation

NPS	150#	300#	600#
1/2"	0.14	0.21	0.28
3/4"	0.14	0.28	0.35
1"	0.14	0.28	0.42
1 1/2"	0.28	0.42	0.56
2"	0.42	0.98	1.12
3"	0.77	1.54	2.38
4"	1.40	2.45	3.43
6"	2.31	4.20	5.81
8"	3.43	4.76	6.44
10"	4.20	6.09	8.26
12"	5.67	6.93	10.15
14"	6.02	7.14	13.58
16"	7.28	10.71	16.80
18"	8.19	11.69	23.80
20"	9.03	14.70	27.30
24"	11.69	21.70	37.10

8.3.1 Installation Recommendations

To properly install a Wafer-type valve, which is designed to be mounted between two flanges, ensuring precise pipe alignment is essential.

The valve must be perfectly centered so that its disc does not strike the pipe when opening.

Before tightening, it is vital to verify that the disc can rotate 90 degrees without obstruction. The long bolts that secure it must be tightened gradually and in a cross-pattern to ensure uniform pressure and prevent damage.

Generally, no additional gaskets are needed, as the valve's own elastic seat acts as a seal.

The flange faces must be checked to ensure they are clean, undamaged, and parallel.

8.4 Installing Butt-Weld End Valves

This Table details the man-hour times for handling and positioning valves with butt-weld ends. It does not include welding times.

The data is classified by the valve's diameter (NPS) and series.

Table 8.3. Man-hours for Butt-Weld End Valve Installation

NPS	150#	300#	600#	900#	1500#	2500#
1/2"	0.21	0.21	0.28	–	0.49	0.63
3/4"	0.21	0.35	0.42	–	0.70	0.77
1"	0.21	0.35	0.42	–	0.77	0.91
1 1/2"	0.35	0.49	0.63	–	1.19	1.40
2"	0.49	1.12	1.33	–	1.40	1.61
3"	0.91	1.75	2.73	3.08	3.43	3.92
4"	1.75	2.80	3.99	4.83	5.67	6.65
6"	2.59	4.83	6.65	7.14	7.56	8.75
8"	3.92	5.53	7.42	8.40	9.38	10.85
10"	4.83	7.00	9.52	10.78	12.04	14.00
12"	6.51	7.98	11.69	14.00	16.08	18.20
14"	6.93	8.19	15.40	–	–	–
16"	8.40	12.32	19.60	–	–	–
18"	9.38	13.44	27.30	–	–	–
20"	10.36	17.50	31.50	–	–	–
24"	13.44	25.20	42.70	–	–	–

8.4.1 Welding Instructions

During the welding of a *butt-weld* valve, the following instructions must be strictly followed:

➔ **Follow manufacturer's instructions:** Strictly adhere to the recommended procedures and verify pipe alignment.

➔ **Valve position:** Keep it partially open to reduce thermal stress and prevent deformation.

➔ **Cleaning and protection:** Clean the pipe ends and protect the valve's interior from spatter and slag, either with covers or inert gas. Prevent the seats and seals from exceeding the maximum allowed temperature.

8.5 Man-Hours for Installing Safety and Control Valve

This Table indicates the man-hour times for the installation of **safety (pressure relief) valves and control valves.**

The data is classified according to the **diameter (NPS) and series (pressure class)** of each valve. It does not include time for bolt torquing.

Table 8.4. Installation of Safety and Control Valves (Man-Hours)

NPS	150# (Mhr)	300# (Mhr)
1/2"	0.84	0.91
3/4"	0.84	1.26
1"	0.84	1.26
1 1/2"	1.05	1.75
2"	1.47	3.15
3"	2.03	4.41
4"	2.73	5.81
6"	5.53	9.38
8"	7.31	10.78
10"	9.87	14.70
12"	12.11	16.80
14"	13.16	18.20
16"	16.10	24.50
18"	18.20	28.00
20"	21.00	32.90
24"	25.90	45.50

8.5.0.1 Safety Valves

For safety valves, the installation must always be **vertical** with the bonnet facing up, as close as possible to the equipment to be protected, and without any intermediate shut-off valves that could inadvertently isolate it.

It is crucial that the inlet piping is short, direct, and has a diameter no smaller than the valve's inlet connection to prevent pressure drops that could affect its performance.

Discharge pipes must have a diameter equal to or greater than the outlet, be adequately supported so as not to exert stress on the valve body, and have drain holes to prevent condensate accumulation.

The discharge must be directed to a safe area, away from personnel and sensitive equipment, ensuring that the discharge does not create additional risks or backpressures that exceed the valve's design limit.

8.5.0.2 Control Valves

For control valves, it is imperative to respect the **flow direction** indicated by an arrow on the body.

They require straight pipe runs, typically 5 to 10 pipe diameters upstream and 3 to 5 diameters downstream, to ensure stable flow and precise control, minimizing turbulence.

It is good practice to install a Y-type or basket strainer upstream to protect the valve's internals from particles and debris carried by the fluid.

Additionally, sufficient space must be provided around the valve to facilitate access for maintenance, calibration, or actuator removal.

They are often installed in a *bypass* configuration with block valves, which allows the flow to be diverted for servicing the control valve without interrupting the process.

Chapter 9

Assembling Pipe Supports

9.1 Objective

To establish standard man-hour (Mhr) estimates for the fabrication and assembly of pipe supports in industrial plants. The values, detailed in Table 9.1, vary according to the nominal pipe size (NPS) and schedule (SCH) of the pipe being supported.

9.1.1 Scope

The time estimates in this chapter apply exclusively to supports that are welded to the pipe. This scope focuses on attachments that become an integral part of the pressure boundary or support system through welding, requiring specific fit-up, inspection, and skilled labor. The following components are covered:

- ➔ **Wear Plates:** Sacrificial elements designed to protect the pipe from frictional wear where it rests on structural supports or is held by non-welded clamps. Their primary function is to prevent localized wall thinning caused by operational vibration or thermal expansion.
- ➔ **Pipe Shoes:** Welded structures that serve as a sliding interface on beams or sleepers. By elevating the line, they prevent direct pipe-to-steel contact which mitigates corrosion and allows for controlled axial movement during thermal cycles.
- ➔ **Lugs:** Welded anchor points for connecting rods, clamps, or other components.
- ➔ **Trunnions:** Sections of pipe welded perpendicularly to the main line (commonly at elbows) to create a rigid support point. They are frequently used for supporting heavy vertical lines or as robust anchors where standard supports are impractical.

Table 9.1. Table for Piping Support Calculation

NPS	Pipe Weight in kg/m		Support Spacing in meters	Support Weight in kg/m		Support Prefabrication (Mhr/m)		Support Installation (Mhr/m)	
	SCH 40	SCH 80		for SCH 40 pipe	for SCH 80 pipe	for SCH 40 pipes	for SCH 80 pipes	on SCH 40 pipes	on SCH 80 pipes
1"	2.50	3.25	2.14	0.50	0.65	0.50	0.65	0.10	0.13
1 1/2"	4.10	5.40	2.75	0.80	1.00	0.60	0.80	0.10	0.15
2"	5.40	7.47	3.05	1.00	1.35	0.70	0.90	0.20	0.22
2 1/2"	8.50	11.50	3.35	1.30	1.70	0.75	1.00	0.20	0.26
3"	11.30	15.25	5.65	1.50	2.00	0.80	1.10	0.20	0.30
4"	16.06	22.29	4.30	2.00	2.70	0.90	1.20	0.27	0.36
6"	28.30	42.51	5.20	3.00	4.70	1.20	1.80	0.40	0.60
8"	33.30	–	5.80	4.00	–	1.50	–	0.50	–
10"	60.30	–	6.70	5.00	–	1.70	–	0.60	–
12"	73.80	–	7.00	6.00	–	1.90	–	0.70	–
14"	81.80	–	7.60	6.60	–	2.00	–	0.70	–
16"	93.10	–	8.25	7.50	–	2.20	–	0.80	–
18"	105.00	–	8.55	8.50	–	2.30	–	0.80	–
20"	117.00	–	9.15	9.50	–	2.40	–	0.90	–
22"	130.00	–	9.50	11.00	–	2.50	–	0.90	–
24"	141.00	–	9.75	12.00	–	2.60	–	1.00	–

9.1.2 Handling

This includes unloading the component at the worksite, storage in the workshop, and subsequent loading and transport to the installation point.

9.1.2.1 The time estimates include:

- ➔ **Handling.**
- ➔ **Support prefabrication.**
- ➔ **Assembly.**

9.1.2.2 The time estimates do not include:

- ➔ **Fabrication and assembly of pipe racks or sleepers.**
- ➔ **Fixed points (anchors).**

- ➔ **Assembly of spring supports.**
- ➔ **Fabrication of auxiliary structures.**
- ➔ **Cleaning and painting of supports.**

9.1.3 ASME B31.3 Considerations for Supports Attached to Pipes

The ASME B31.3 standard establishes specific requirements for the fabrication and assembly of supports that are attached directly to the pipe to avoid compromising its integrity. Key considerations for different components are summarized below:

- ➔ **Wear Plates, Trunnions, and Pipe Shoes:** For components welded directly to the pipe, the fundamental consideration is **material compatibility and welding qualification (WPS/WPQ)**. It is crucial that the support material is compatible with the pipe material (typically identical or from the same P-Number family) to ensure a proper metallurgical bond and prevent galvanic corrosion.
- ➔ **Welded Lugs:** For these elements, not only must the welding be qualified, but the component design is also critical to **avoid stress concentrations**. For example, a lug must be designed with a wide base and smooth edges so that the load from the U-bolt or clamp is distributed evenly over a larger area on the pipe wall.

9.1.4 Preventing Galvanic Corrosion

The fundamental principle for preventing galvanic corrosion is to **avoid direct contact between dissimilar metals**, such as stainless steel and carbon steel. This contact, in the presence of moisture, creates a galvanic couple that accelerates the corrosion of the less noble metal.

The correct practice to protect the integrity of the pipe is to install a physical barrier. For example, when securing a stainless steel pipe with a carbon steel clamp, it is mandatory to insert a **wear plate**. Crucially, this plate must be made of the **same material as the pipe** (stainless steel) and welded to it, ensuring that the pipe is electrically isolated from the clamp.

Chapter 10

Productivity for Pipe Cleaning and Painting

10.1 Objective

The objective of this chapter is to calculate the required man-hours for the cleaning and application of a standard paint system on carbon steel pipe surfaces. The productivity rates presented correspond to the following specific conditions:

- ➔ Environmental Corrosivity Level: C3 (Medium), according to the ISO 12944 standard (See Table 10.1).
- ➔ Application Method: The application is primarily done with spray equipment to ensure uniform coverage of the primer and finish coats, supplemented with manual application (roller and brush) for touch-ups.
- ➔ Surface Preparation: Grade Sa 2½ (near-white blast cleaning) or its equivalent, SP-10, according to ISO 8501 / SSPC standards. This preparation provides a suitable anchor profile for the first coat of primer.

10.1.1 Application Conditions

The productivity rates in Table 10.2 are valid for medium-sized work batches. For smaller-scale jobs or spot repairs, specific rates should be used, such as those illustrated in the practical examples included in this manual.

The specified Dry Film Thickness (DFT) must be guaranteed.

Table 10.1. Classification of Environments by Corrosivity Level.

Category	Corrosivity Level	Typical Environment Description
C1	Very Low	Heated buildings.
C2	Low	Atmospheres with low pollution; mainly rural areas.
C3	Medium	Urban and industrial environments, with moderate pollution; coastal areas with low salinity.
C4	High	Industrial areas and coastal zones with moderate salinity.
C5-I	Very High (Industrial)	Industrial areas with high humidity and aggressive atmospheres.
C5-M	Very High (Marine)	Coastal and offshore areas with high salinity.

Table 10.2. Performance Rates for Surface Treatment and Painting of Piping.

NPS	External Surface (m^2/m)	Total Paint & Clean SP-10 (Mhr/m)
1"	0.10	0.28
1 1/2"	0.15	0.35
2"	0.19	0.41
2 1/2"	0.23	0.48
3"	0.28	0.65
4"	0.36	0.82
6"	0.53	1.00
8"	0.69	1.10
10"	0.85	1.21
12"	0.96	1.38
14"	1.12	1.56
16"	1.28	1.73
18"	1.44	2.08
20"	1.60	2.59
24"	1.91	2.08
30"	2.39	2.59

10.1.2 Equivalent Lengths of Fittings

For man-hour estimation in surface treatment and painting, each fitting must be converted to its equivalent length of straight pipe according to the following criteria:

➔ **NPS $\leq$ 4:** Uniform equivalent length of 1.0 lm for all fittings.

➔ **NPS > 4:** Apply values from Table 10.3 according to fitting type.

Table 10.3. Equivalent lengths of fittings for painting and cleaning estimation (NPS > 4).

Fitting Type	Equivalence (lm)	Considerations
90° Elbow / Tee	1.5 – 2.0	Greater surface area and geometric complexity
45° Elbow / Reducer	1.0	Less complex geometry
Flange	2.5 – 3.0	Outer surface, hub, and bolting. Sealing faces are masked*
Valves	2.0	Finish coat only**

***Flange faces:** Sealing surfaces (raised face, RTJ, etc.) are masked with protective materials during sandblasting and painting processes to preserve the integrity of the flange faces.

****Valves:** Supplied with shop primer; require cleaning and finish coat only.

General note: The values indicated represent linear meters (lm) of equivalent straight pipe that must be added to the total piping length for man-hour calculation.

10.1.3 Scope and Parameters of Table 10.2

Tasks Included in Performance Rates:

➔ Material Handling: Unloading, laydown yard storage, and transportation of piping materials.

➔ Quality Control: Time for inspection of applied coating (e.g., thickness measurements).

➔ Touch-ups and quality control on installed piping.

10.1.4 Considerations for Paint Systems

A typical industrial paint system for carbon steel consists of three coats:

Primer This is the first coat, applied directly to the prepared steel. Its function is to provide excellent adhesion and active anti-corrosion protection (e.g., zinc-rich primers).

Intermediate Coat Applied over the primer to increase the total system thickness (Dry Film Thickness or DFT). It acts as a physical barrier against the ingress of water and oxygen.

Finish Coat (Top Coat) This is the visible layer. It provides the final color and protection against environmental elements, such as UV radiation (which degrades the underlying coats), abrasion, and chemical spills.

10.1.5 Inspection and Quality Control

Throughout the process, inspections are conducted at critical stages to monitor surface cleanliness for proper adhesion, anchor profile for mechanical bonding, and environmental application conditions.

The Dry Film Thickness (DFT) of each coat is also meticulously verified, as insufficient thickness compromises barrier protection while excessive application can lead to cracking or improper curing. In high-performance systems, a *holiday* test (porosity detection) may also be performed to ensure the absolute continuity of the paint film, which is the final quality check for creating a truly impermeable protective barrier..

Chapter 11

Installing Welding Neck Flanges

11.1 Objective and General Scope

The objective of this chapter is to establish the man-hour rates for the assembly of Welding Neck flanges.

The values presented in this chapter's table apply to carbon steel flanges (ASTM A105) installed in above-ground plants or facilities.

11.1.1 Correspondence Between Flange Class and Pipe Schedule

- ➔ **150#** to **600#** Class for **STD** pipe.
- ➔ **900#** to **1500#** Class for **SCH 80** pipe.
- ➔ **2500#** Class for **SCH 160** pipe.

11.1.2 Key Definitions

Handling: This comprises the complete component handling cycle on-site: unloading, initial storage in the workshop or laydown area, and subsequent transport to the exact point of assembly.

Rating (Pressure Class): Flanges are categorized into pressure classes (commonly called Ratings and designated with #). This classification indicates their ability to withstand different combinations of pressure and temperature.

Table 11.1. Weight (kg) and Man-hour (Mhr) Specifications by NPS and Pressure Class

NPS	150#		300#		600#		900#		1500#		2500#	
	kg	Mhr	kg	Mhr	kg	Mhr	kg	Mhr	kg	Mhr	kg	Mhr
1/2"	0.80	1.30	0.91	1.40	1.40	1.60	3.00	1.70	3.10	1.80	3.70	2.00
3/4"	0.90	1.40	1.36	1.50	1.80	1.60	3.20	1.80	3.30	1.90	4.10	2.20
1"	1.15	1.50	1.80	1.70	2.30	1.70	3.90	1.90	3.90	2.20	5.90	2.40
1 1/2"	1.90	1.80	3.20	2.00	4.50	2.10	6.40	2.30	6.60	2.80	12.70	3.50
2"	2.70	2.20	3.60	2.40	5.40	2.60	10.90	2.80	10.90	3.40	19.00	4.20
2 1/2"	4.60	2.50	5.40	2.90	8.20	3.30	14.50	3.40	16.30	4.50	24.00	5.00
3"	5.20	3.10	7.30	3.40	10.40	3.80	16.30	4.30	21.80	5.20	42.70	6.00
4"	7.30	3.60	11.30	3.90	19.10	4.60	23.00	5.10	31.30	6.30	66.30	8.00
6"	10.90	4.70	20.40	5.40	36.70	6.20	50.00	7.00	74.40	9.00	171.00	13.00
8"	19.10	6.00	31.80	7.00	53.10	8.20	85.00	9.20	123.90	12.50	261.00	20.00
10"	25.00	7.40	44.90	9.20	85.70	10.30	121.60	12.10	206.00	16.50	485.00	30.00
12"	39.00	8.80	64.40	11.00	102.50	12.60	168.70	14.50	313.00	23.00	729.00	40.00
14"	52.00	10.50	84.40	13.00	157.40	14.50	255.00	16.70	–	–	–	–
16"	65.00	12.20	111.60	14.00	218.00	17.00	311.00	18.50	–	–	–	–
18"	70.30	14.00	138.40	16.50	252.00	19.20	419.00	21.00	–	–	–	–
20"	77.00	16.00	171.50	18.50	313.00	23.00	528.00	24.50	–	–	–	–
24"	118.00	18.00	247.20	21.00	444.00	30.00	956.00	35.00	–	–	–	–

Standard classes are 150, 300, 400, 600, 900, 1500, and 2500. The higher the class, the more robust the flange. A flange's maximum working pressure decreases as the service temperature increases.

11.1.3 References and Standards

ASME B16.5: Covers specifications for pipe flanges from NPS 1/2 through 24 inches. **ASME B16.47:** Covers specifications for large diameter flanges from NPS 26 through 60 inches.

(Note: The flange weights used in this manual have been sourced from the Tube-Turn manufacturer's catalog.)

11.1.4 Scope of Table Values

Included Tasks:

➔ Component handling.

➔ Assembly and alignment of the flange to the pipe.

- ➔ Welding of the butt-weld joint.
- ➔ Installation of bolts and gasket (without tightening).

Explicitly Excluded Concepts:

- ➔ Final tightening or tensioning of bolts.
- ➔ Fabrication and installation of supports.
- ➔ Cleaning and painting work.
- ➔ Application of thermal insulation.

11.2 Overview of the Gasket-Flange-Bolt System

The most common cause of leaks in flanged joints is an improper installation procedure.

11.2.1 Visual Inspection Before Assembly

11.2.1.1 Inspection Recommendations

- ➔ Inspect the flange seating surface for defects such as burrs, damage, corrosion, etc., to take corrective action.
- ➔ Defects that cross the flange face radially are especially concerning. Notify a supervisor of any questionable imperfections so they can be addressed properly.
- ➔ Clean the seating surfaces with soft brushes to avoid damaging the existing surface finish.
- ➔ Brushing should always be done in the direction of the serrations (not perpendicular to them) to minimize wear.

11.3 ASME PCC-1 Tolerances

Correct alignment of all components in the flanged joint ensures maximum contact of the seating surface and a uniform load on the gasket, thereby improving the effectiveness of all fastener tightening methods.

11.3.1 Parallelism

Flange parallelism is defined as the alignment of pipe or vessel flanges so that there are equal distances between the flange faces at all points around the joint's circumference. Verification of parallelism is done by measuring the minimum and maximum gap between the flange faces and comparing the two values to detect deviations. An acceptable deviation is no more than 0.8 mm (1/32 in). (see figure).

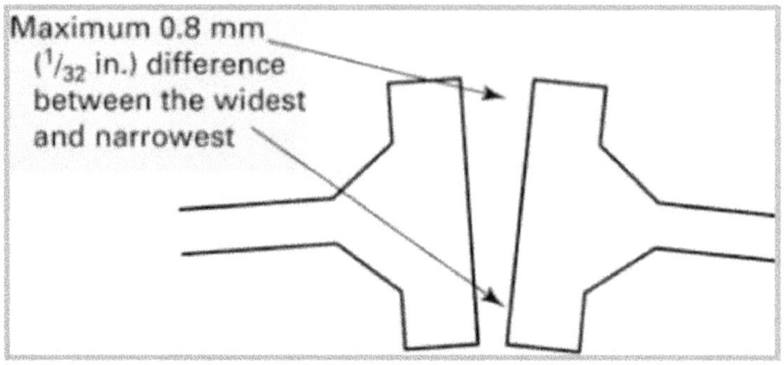

Figure 11.1. Flange Parallelism Tolerance

11.3.2 Rotational Alignment

Rotational alignment is defined as the height difference between the bolt holes located on opposite sides of the flange centerline. This difference is measured using a level or a height gauge, referencing a flat, fixed surface (see figure).

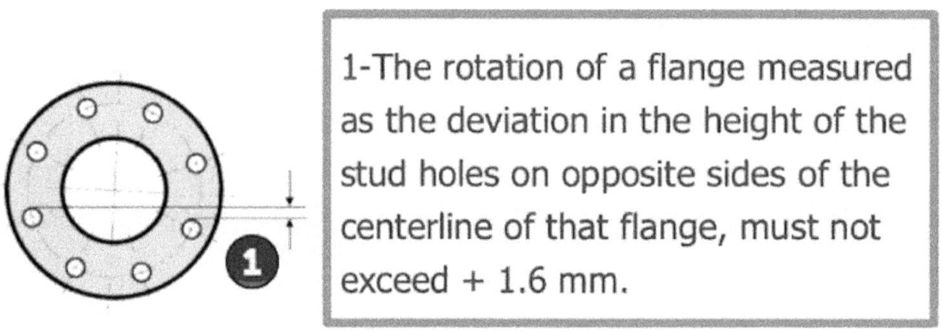

Figure 11.2. Rotational Tolerance.

11.3.3 Gap

An excessive space or separation between a flange and its mating flange.

- ➔ It must be verified that, even while meeting the maximum allowable deviations for parallelism and rotation, the gap between the flanges at rest does not exceed twice the gasket thickness.
- ➔ When the flange gap is excessive, the force applied to bring them together should not exceed 10% of the maximum torque required for proper joint tightening.

11.4 Flange Alignment Guidelines

1. Out-of-tolerance conditions, such as excessive flange gap or rotational misalignment, must be corrected before installing the gasket. A compromised gasket is the primary cause of joint leakage; forcing it into a misaligned joint can cause crushing or extrusion, leading to immediate or future failure. Only minimal adjustments are permitted after the gasket is installed.
2. If more force is required for alignment than can be exerted by hand or with common tools such as spud wrenches, consult an expert. This indicates a systemic issue like improper support or fabrication errors. Applying excessive mechanical force is a dangerous practice that introduces high residual stress into the piping system, potentially leading to premature weld failure.
3. Proper alignment will allow the bolts to pass freely through the flanges at a right angle and the nuts to rest flat against the flanges before tightening. This ensures a uniform bolt load distribution upon tightening, which is critical for achieving a reliable seal across the entire gasket face.
4. Before using jacks, chain hoists, or other mechanical devices to force alignment, it is advisable to perform a pipe stress analysis. This is especially critical if the piping is old or if walls are suspected to have thinned from corrosion, as the applied forces could exceed the pipe's mechanical integrity and cause damage.
5. If the flanges needing alignment are connected to pumps or rotating equipment, great care must be taken. The nozzles on such equipment have strict load limits; exceeding them can cause shaft misalignment, leading to accelerated bearing wear, seal failure, and potentially catastrophic damage to the internal rotating assembly.
6. The best practice for repairing a misaligned component is to remove it and reinstall it correctly. While it may seem time-consuming, this is the only method that ensures a stress-free connection. Forcing a joint only hides the underlying problem and transfers dangerous stress elsewhere in the system.

Chapter 12

Installing Slip-On Flanges

12.1 Objective and General Scope

The purpose of this chapter is to detail the man-hour rates required for the assembly of Slip-On flanges.

The values presented in Table 12.1 correspond to carbon steel flanges (ASTM A105) installed in at-grade facilities.

12.1.1 Correspondence Between Flange Class and Pipe Schedule

The correspondence between the flange pressure class and the pipe schedule is as follows:

- ➔ **150#** to **600#** Class for **STD** pipe.
- ➔ **900#** to **1500#** Class for **SCH 80** pipe.
- ➔ **2500#** Class for **SCH 160** pipe.

12.1.2 Definitions

Handling: This comprises the complete on-site handling cycle of the component: unloading, initial storage in the workshop or laydown area, and subsequent transport to the exact point of assembly.

Rating or Pressure Class: Flanges are categorized into pressure classes (commonly called Ratings and designated with #). This classification indicates their ability to withstand different combinations of pressure and temperature. Standard classes are 150, 300, 400, 600, 900, 1500, and 2500. The higher the class, the more robust the flange.

Table 12.1. Adjusted Specifications with Abbreviated Titles (Mhr)

NPS	150#		300#		600#		900#		1500#		2500#	
	kg	Mhr	kg	Mhr	kg	Mhr	kg	Mhr	kg	Mhr	kg	Mhr
1/2"	0.45	1.00	0.70	1.10	0.90	1.25	2.70	1.40	–	–	3.30	1.60
3/4"	0.70	1.10	1.20	1.20	1.40	1.30	2.80	1.44	–	–	4.10	1.80
1"	0.90	1.20	1.36	1.40	1.60	1.40	3.40	1.50	–	–	5.50	1.90
1 1/2"	1.40	1.45	2.80	1.60	2.90	1.70	6.40	1.80	–	–	12.00	2.80
2"	2.30	1.80	3.20	1.90	3.70	2.10	11.40	2.20	–	–	17.30	3.40
2 1/2"	3.63	2.00	4.54	2.30	5.50	2.60	16.30	2.50	–	–	24.90	4.00
3"	4.08	2.48	5.90	2.70	6.80	3.00	14.00	3.50	21.70	4.20	37.60	4.80
4"	5.44	2.90	9.60	3.10	15.00	3.70	24.00	4.10	33.10	5.10	57.60	6.40
6"	8.20	3.80	15.90	4.30	36.00	5.00	49.00	5.60	74.40	7.20	147	10.00
8"	12.70	4.80	24.50	5.60	44.00	6.60	78.00	7.40	117	10.00	220	16.00
10"	16.80	6.00	35.00	7.40	80.00	8.20	111	9.70	198	13.00	420	24.00
12"	27.00	7.00	50.00	8.80	98.00	10.00	148	12.00	303	18.00	590	32.00
14"	35.00	8.40	75.00	10.00	117	12.00	172	13.00	–	–	–	–
16"	42.00	9.80	100	11.00	166	14.00	208	15.00	–	–	–	–
18"	55.00	11.20	127	13.20	216	15.40	293	16.80	–	–	–	–
20"	71.00	13.00	147	15.00	278	18.00	359	19.60	–	–	–	–
24"	96.00	14.00	222	17.00	397	24.00	671	28.00	–	–	–	–

12.1.3 Scope of Table Values

Included Tasks:

- ➔ Handling, assembly, and alignment of the flange with the pipe.
- ➔ Execution of the two fillet welds (internal and external).
- ➔ Installation of bolts and gasket (without tightening).

Explicitly Excluded Concepts:

- ➔ Final tightening or tensioning of bolts.
- ➔ Fabrication and installation of supports.
- ➔ Cleaning and painting work.
- ➔ Application of thermal insulation.

Chapter 13

Estimating Welding Consumables

13.1 Resource Estimation for SMAW Welding

To facilitate planning and cost estimation, the following table presents the necessary resources for butt weld joints in carbon steel piping. The data is based on the SMAW process and assumes a deposition efficiency of 55%.

13.1.1 Considerations

This table is a key tool for cost engineering and planning of piping projects, as it allows for an accurate estimation of resources. Its parameters are defined as follows:

NPS (Nominal Pipe Size): Nominal diameter of the pipe.

SCH (Schedule): The wall thickness of the pipe. A higher SCH value implies a thicker wall and, therefore, greater resource consumption.

gr (Grams per joint): Weight of the filler material (electrodes) required per weld, an essential data point for materials management.

Mhr (Man-hours per joint): Estimated work time per joint, used to calculate labor costs and define the project schedule.

13.1.2 Welding Consumables per 100 kg of Deposited Metal by Process

The following table presents the consumables used in different welding processes to achieve a deposition equivalent to 100 kg of metal.

Table 13.1. Electrode Consumption and Man-hours (Mhr) for Butt Joints (SMAW Process)

NPS	SCH 10		SCH 20		SCH 30		SCH 40		SCH 60		SCH 80		SCH 160	
	g	Mhr	g	Mhr	g	Mhr	g	Mhr	g	Mhr	g	Mhr	g	Mhr
2"	–	–	–	–	–	–	180	1.50	–	–	260	1.70	400	2.00
2 1/2"	–	–	–	–	–	–	210	1.80	–	–	300	2.10	480	2.50
3"	–	–	–	–	–	–	280	2.00	–	–	380	2.50	550	3.00
4"	–	–	–	–	–	–	360	2.50	–	–	550	3.00	700	4.00
6"	–	–	–	–	–	–	520	3.50	–	–	900	4.50	1400	6.50
8"	–	–	430	4.00	500	4.50	700	4.50	–	–	1200	6.20	2400	10.50
10"	–	–	540	4.50	680	5.00	850	5.80	–	–	1500	8.50	2800	16.00
12"	–	–	650	5.00	850	6.00	1300	7.00	–	–	2250	11.50	4200	22.50
16"	840	5.80	1120	7.00	1460	8.00	2350	10.50	–	–	–	–	–	–
18"	940	6.40	1380	8.20	2000	10.00	–	–	–	–	–	–	–	–
20"	1050	6.80	1800	9.50	2350	13.00	–	–	–	–	–	–	–	–
22"	1180	7.20	2050	10.50	–	–	–	–	–	–	–	–	–	–
24"	1300	8.00	2250	11.50	–	–	–	–	–	–	–	–	–	–
30"	2750	10.00	3700	14.00	–	–	–	–	–	–	–	–	–	–
32"	–	–	–	–	–	–	–	–	–	–	–	–	–	–
36"	–	–	–	–	–	–	–	–	–	–	–	–	–	–

Table 13.2. Consumables per 100 kg of Deposited Metal

Process	Consumables / 100 kg Deposited Metal		
	Electrode (kg)	Flux (kg)	Gas (m^3)
Cellulose-coated manual electrode	155	—	—
Rutile-coated manual electrode	145–170	—	—
Low-hydrogen manual electrode	160–170	—	—
MIG (short circuit)	110	—	17–42
MIG (spray)	108	—	7–11
Shielded flux-cored	122	—	4–20
Self-shielded flux-cored	126	—	—
Submerged arc	102	85–100	—

Chapter 14

Productivity for Cleaning and Painting Structures

14.1 On-Site Treatment Phases for Metal Structures

1. **Unloading and Initial Staging** Safe reception and unloading of the structure using cranes or forklifts. Materials are organized in a designated area and checked against the delivery ticket or drawing.
2. **Transfer to the Work Area:** Internal movement of profiles from the staging area to the treatment area, utilizing cranes and dollies.
3. **Area Protection and Containment:** The work area is contained to control dust and paint overspray, which protects nearby personnel and equipment.
4. **Equipment Preparation:** Machinery is prepared for each phase. For abrasive blasting, compressors and hoppers are connected. For painting, the equipment is set up, and the product is mixed according to specifications.
5. **Handling and Positioning of Profiles:** Placing and rotating materials on sawhorses or trestles to expose all surfaces for treatment.
6. **Abrasive Blasting (Surface Preparation):** High-pressure projection of an abrasive to remove rust, thereby achieving the required degree of cleanliness and anchor profile for paint adhesion.
7. **Stripe Coating:** Mandatory application of a paint coat with a brush on edges, corners, and vertices to ensure protection in these vulnerable areas.

8. **Application of Paint Coats:** Uniform spraying of primer and finish coats until the specified Dry Film Thickness (DFT) is achieved, guaranteeing complete anti-corrosion protection.
9. **Quality Control and Inspection:** Final verification of the treatment, including thickness measurements and a visual review to detect any potential defects.

14.2 Coefficients for Estimating Abrasive Blasting and Painting

To calculate the total area to be painted on support structures, the following standard coefficients, which relate surface area (m^2) to weight (tons), can be used:

- ➔ Heavy Structures: 25 m^2/ton
- ➔ Light Structures: 35 m^2/ton

14.3 Basis and Scope of Productivity Rates

The man-hour values presented in the following table correspond to a paint system designed for a C3 (Medium) environmental corrosivity level, according to the ISO 12944 standard.

14.4 Basis and Scope of the Man-Hour Table for Painting

The man-hour values for cleaning and coating application presented in the performance rate table below correspond to a painting system. It is designed for a C3 (medium) environmental corrosivity level according to ISO 12944.

Man-hours per Ton

Table 14.1. Performance Rates for Cleaning and Painting of Structures

Structure Type	m^2/ton	Mhr/ton
Heavy structure	25	38
Light structure	35	70

Chapter 15

Practical Example 1: Piping System Assembly

15.1 Scenario Description and Methodology

This chapter breaks down a comprehensive case study for calculating direct labor, focusing on the following key characteristics that define its complexity and practical relevance:

➔ **Practical Objective:** To illustrate in detail the estimation methodology in a realistic project involving the replacement of a piping line, providing a robust calculation model applicable to similar industrial situations.

➔ **Challenging Environment (Brownfield):** All activities are carried out within a fully operational industrial plant. This introduces operational and safety complexities that do not exist in new (greenfield) projects and serves as an excellent example for risk analysis and planning.

➔ **Focus on Adjustment Factors:** It analyzes how site constraints (permits, LOTO systems, interferences) impact productivity. Mastering the quantification of these variables is fundamental for an accurate estimation, avoiding cost and schedule deviations.

➔ **Favorable Contrasting Condition:** The work is performed in a covered area with a concrete floor at ground level. This condition simplifies logistics and eliminates the impact of weather, allowing other productivity variables to be isolated and analyzed more clearly.

15.1.1 Estimation Process

The calculation will be performed in two stages:

➔ **Base Calculation:** First, the Standard Tabulated Hours (STH) will be determined. These represent the initial reference value for estimating direct labor.

  These hours come directly from the tables included in this manual, developed from actual site data collected and applied by the Author throughout his professional career, which guarantees their practical backing and relevance in real construction environments.

➔ **Adjustment for Real Conditions:** Subsequently, the STH value will be modified by applying the adjustment factors detailed in Chapter 3.

  This process aims to adapt the estimate to the specific conditions of the project, such as performing work in an active industrial plant with its associated limitations and risks.

The piping to be analyzed is the section between weld points 1 and 10, as shown in the following figure.

Figure 15.1. Piping to be replaced.

Table 15.1. Bill of Materials for Piping

Item	Qty	Description	Material and Specification
1	0.80 m	Pipe NPS 4"	Carbon Steel, SCH 40, ASTM A53
2	3	90° Elbow, Long Radius (LR)	Carbon Steel, ASTM A234 Gr. B, SCH 40
3	2	Slip-On Flange, NPS 4"	Carbon Steel, Class 150, ASTM A105
4	1	Gate Valve, flanged, NPS 4"	Class 150

15.1.2 Bill of Materials (According to Table 15.1)

15.2 Phase 1: Calculating Standard Tabulated Hours (STH)

In this phase, the base hours are calculated using the values from the manual's tables, assuming standard conditions.

15.2.1 Step 1: Calculating Material Weight

Using the bill of materials and technical reference tables, the weight of the piping is calculated as shown in Table 15.2

Table 15.2. Material Weight Calculation

NPS	Material	Unit Weight	Quantity	Total Weight per item	Reference Table
4"	90° LR Elbow SCH 40	3.86	3.00	11.58	5.1
4"	A53 Pipe SCH 40	16.06	0.80	12.85	4.1
4"	SO Flange S150#	5.44	2.00	10.90	12.1
4"	Gate Valve S150#	64.00	1.00	64.00	8.1
		Total weight in kg		99.33	

15.2.2 Step 2: Calculating Man-Hours for Supports

Calculation of equivalent length: This is obtained by dividing the total weight of the piping by the unit weight of the representative pipe, in this case, an NPS 4" SCH 40 pipe. That is:

$$\frac{99.33 \text{ kg}}{16.06 \text{ kg/m}} = 6.19 \text{ m}$$

Table 15.3. Man-Hour Calculation for NPS 4" Pipe Supports

NPS	Equiv. Length (A)	Support Weight	Fabric., and Assembly Man-hours per meter of pipe (B)	Total Mhr (A*B)	Ref. Table
4"	6.19	2.00	1.17	7.24	9.1

Man-hour calculation: Using the Table with the equivalent length of 6.19 m, the hours for supports are obtained.

Result: The total hours for supports is $6.19 \times 1.17 = 7.24$ Mhr.

15.2.3 Step 3: Estimating Man-Hours for Pipe Cleaning and Painting

The equivalent length to be painted is equal to the pipe length plus one meter for each fitting.

15.2.3.1 Calculation of the length to be painted:

$$0.80\,\text{m (straight run)} + (6\,\text{ea} \times 1\,\text{m/ea}) = 6.80\,\text{m}$$

(Note: 3 elbows, 2 flanges, and 1 valve are added = 6 units)

15.2.3.2 Calculation of the total area to be painted:

$$6.80\,\text{m} \times 0.36\,\text{m}^2/\text{m (from table for NPS 4")} = 2.45\,\text{m}^2$$

Man-hour calculation: An Mhr/m² ratio for small batches is applied. **Result:** The Total hours for

Table 15.4. Summary of Man-Hours for Abrasive Blasting and Painting

Activity	Man-hours per m^2	Surface Area to be treated	Total per item	Reference Table
Abrasive Blasting	1.50	2.45	3.68	N/A
Painting and touch-up	2.20	2.45	5.39	N/A
Total Mhr for blasting and painting			9.07	

piping painting are **9.07 Mhr**.

15.2.4 Step 4: Calculating Man-Hours for Support Cleaning and Painting

Calculation of support weight:

$$6.19\,\text{m (equiv. length)} \times 2\,\text{kg/m (from table)} = 12.38\,\text{kg}$$

Calculation of area to be painted:

$$0.01238\,\text{Tons} \times 35\,\text{m}^2/\text{Ton (light structures ratio)} = 0.43\,\text{m}^2$$

Man-hour calculation: The same Mhr/m² ratio for small batches as in the previous step is applied. **Result:** The total hours for painting supports is 1.60 Mhr.

Table 15.5. Man-Hour Calculation for Cleaning and Painting of Supports

Activity	Kg of support (A)	Area per Ton m² * Ton (B)	Surface Area per item A*B=C	Mhr/m² (D)	Total per item C*D=E
Abrasive Blasting	12.38	35.00	0.43	1.50	0.65
Painting	12.38	35.00	0.43	2.20	0.95
				Total man-hours	1.60

15.2.5 Step 5: Calculating STH for Assembly and Welding (No Prefabrication)

The man-hours for each component (joints, fittings, etc.) are summed according to the values indicated in the reference tables of Table 15.6.

Table 15.6. Man-Hour Calculation for Assembly and Welding without Prefabrication

NPS	Item	Quantity	Mhr	Total Mhr	Reference Table
4"	90° LR Elbow	3.00	5.00	15.00	5.1
4"	Weld nipple	2.00	Included in fittings	N/A	
4"	Gate valve S150#	1.00	2.00	2.00	8.1
4"	SO Flange S150#	2.00	2.90	5.80	12.1
			Total Mhr	22.80	

Result: The total STH for on-site assembly is 22.80 Mhr.

15.2.6 Step 6: Calculating Mhr with Workshop Prefabrication

In this scenario, fieldwork is drastically minimized, limited only to the two final assembly welds required to connect the system (welds 1 and 10).

The remaining seven joints are executed in a workshop. In this controlled environment, factors such as optimal ergonomics for welders, the absence of weather interference, and the immediate availability of tools allow for much higher productivity than that achieved on-site. For this reason, a **50% reduction** is applied to the tabulated man-hours for these prefabricated joints.

Table 15.7. STH Calculation with Prefabrication

NPS	Item	Quantity	Mhr	Total Mhr	Reference Table
4"	Shop-fabricated joints	7.00	1.25 (2.50/2)	8.75	Table 5.1
4"	Field joints	2.00	2.90	5.80	Table 5.1
4"	Valve, flanged gate S150#	1.00	2.00	2.00	Table 8.1
			Total Mhr:	**16.55**	

15.2.7 Man-Hour Comparison (With vs. Without Prefabrication)

Table 15.8. Comparative Summary of Man-Hours (STH)

Description	Mhr for Complete On-Site Welding	Mhr with Prefabrication
Piping assembly and welding	22.80	16.55
Support fabrication and assembly	7.24	7.24
Piping painting	9.07	9.07
Support painting	1.60	1.06
Totals	**40.71**	**34.46**

15.2.8 Phase 1 Conclusion: STH Summary

The prefabrication strategy demonstrates its efficiency with a net saving of **6.25 man-hours** under standard conditions. This figure, which arises from the difference between the full on-site assembly (40.71 Mhr) and the approach with prefabrication (34.46 Mhr), is equivalent to a reduction of more than 15% in direct labor.

15.3 Phase 2: Applying Adjustment Factors

Now, the STH are corrected to align with the actual project conditions.

15.3.1 Correction A: Project Variable Analysis

The 6 main variables are analyzed to determine if performance will be equal to, lower than, or higher than the standard.

15.3.1.1 Variable 1: Contractor Experience

Evaluation The technical capability and track record of the team are decisive. A team with extensive experience can maintain standard performance, while an inexperienced one will require a negative correction factor. In this case, it is assumed that the installation company is fully familiar with the task.

Assigned Factor 1.00 (Standard Performance).

15.3.1.2 Variable 2: Quality of Information

Evaluation The precision of the engineering is fundamental. Clear drawings and specifications minimize rework. A lack of clarity, on the other hand, requires an increase in the budgeted hours. For this simple task, optimal quality information is assumed.

Assigned Factor 1.00 (Standard Performance).

15.3.1.3 Variable 3: Labor and Union Environment

Evaluation A stable labor environment allows for a continuous workflow. In small projects, union impact is usually minimal. On large-scale projects, it could slow down progress.

Assigned Factor 1.00 (Standard Performance).

15.3.1.4 Variable 4: Site Conditions and Access

Evaluation A covered work area with a level floor and clear access promotes optimal performance. Access difficulties or uneven terrain would necessitate considering a corrective factor.

Assigned Factor 1.00 (Standard Performance).

15.3.1.5 Variable 5: Weather Conditions

Evaluation Weather is a high-impact factor in outdoor work. In this case, the work is performed in an enclosed space, with no weather impact.

Assigned Factor 1.00 (Standard Performance).

15.3.1.6 Variable 6: Work at Height

Evaluation Working at height involves greater risks and lower productivity. For this project, tasks are executed at ground level under standard conditions.

Assigned Factor 1.00 (Standard Performance).

Conclusion for Correction A: The combined adjustment factor for these variables is 1.00. The man-hour consumption is not modified by these concepts.

Correction B: Site Location & Operating Environment

Geographic Location Factor

Not applicable. The project is executed in a standard urban/industrial environment with normal accessibility; therefore, no adjustment is required for extreme geographic conditions (remoteness, high altitude, polar climates, jungle, or desert).

Correction for Congestion & Restricted Access (Brownfield)

Factor Evaluation: This adjustment applies to revamp works performed within operating industrial facilities (*brownfield*) to quantify labor productivity loss resulting from operational interferences. Primary causes include Permit-to-Work administration (*PTW*, *LOTO*), restricted site logistics, Simultaneous Operations (*SIMOPS*) coordination, and specialized PPE requirements.

For this specific case, while the magnitude of this factor might appear minor in the overall estimate, its proper application across any industrial project is critical to prevent budget overruns in live operating environments.

Assigned Factor: An additional **+5 %** (multiplier $1.05\times$) is applied to total direct baseline man-hours.

15.3.2 Correction C: Material and WPS Requirements

Evaluation The materials for the piping are of similar qualities to those used for the data logging of the tables.

Assigned Factor 1.00 (No change).

15.4 Phase 3: Final Estimated Hour Calculation

The only relevant correction factor (+5%) is applied to the STH totals for both options.

Table 15.9. Final Man-Hour Calculation with Adjustment Factors

Description	Mhr for Complete On-Site Welding	Mhr with Prefabrication
Hours required per table	40.71	34.36
Additional for work site location 5%	2.04	1.72
Totals	**42.75**	**36.18**

Chapter 16

Practical Example 2: Labor for Piping Installations

16.1 Project Description

This case study aims to estimate the man-hour (Mhr) consumption required to execute a modification in an industrial plant. The project consists of installing a new centrifugal pump with its suction and discharge piping, connecting it to two existing 6-inch diameter headers.

Figure 16.1. Suction Piping from the Existing Header to the Centrifugal Pump

Figure 16.2. Discharge Piping from the Centrifugal Pump to the Header

16.2 Contractor's Scope of Work

This man-hour estimate precisely details the scope of piping work to be executed by the contractor, establishing a clear framework to avoid deviations. The estimate exclusively covers the following tasks:

- ➔ **Execution of Tie-ins:** The completion of two critical connection points (tie-ins) on the existing headers. This operation involves cutting, preparing bevels, and welding two branch connections using a 6"x4" reducing tee, thereby integrating the new circuit into the current system.
- ➔ **Piping Assembly:** The complete installation and assembly of the new piping, which includes the prefabricated suction and discharge piping, their correct alignment with the pump nozzles and connection points, and the final welding of all components.
- ➔ **Important Exclusion:** As a key delineation of responsibilities, the estimate does not include the labor for the mechanical assembly of the centrifugal pump on its baseplate, nor its subsequent anchoring and laser alignment, as these tasks fall under a different specialty.

16.3 Client's Responsibilities

Certain tasks are the exclusive responsibility of the Client, as they involve aspects that the contractor cannot and should not manage. These are based on three key pillars that ensure the integrity of the operation:

- ➔ **Operational Control:** Only the Client has the authority and comprehensive control to manage the plant. This includes the power to stop and restart production lines, coordinate work that may interfere with other areas, and manage overall site risks, ensuring business continuity.
- ➔ **Personnel and Process Safety:** The Client is the ultimate guarantor of safety, possessing in-depth knowledge of the hazards inherent in its processes (chemicals, pressures, temperatures). It is their responsibility to establish a safe work perimeter and protect the integrity of all personnel, both their own and external contractors.
- ➔ **System Knowledge:** The Client has a deep understanding of existing processes and their interdependencies, allowing them to foresee the consequences of each intervention. Delegating this supervision to a contractor focused only on execution would represent an unacceptable risk to the assets.

16.3.1 Bill of Materials for Suction and Discharge Lines

Table 16.1. Bill of Materials for the Suction Line

Bill of Materials for the Suction Piping	
Description	**Quantity**
Eccentric reducer 4"x3"	1.00
Reducing tee 6"x4"	1.00
Slip-On Flange NPS 4" S150#	2.00
Slip-On Flange NPS 3" S150#	1.00
Gate valve NPS 4" S150#	1.00
Strainer NPS 4" Class 150#	1.00
Elbow NPS 4"	1.00
Pipe 4" NPS	1.50

Table 16.2. Bill of Materials for the Discharge Piping

Description	**Quantity**
Pipe diameter 4" A53 SCH 40	2.50
90° LR Elbow NPS 4", butt weld ends A234 Gr B Std	2.00
Slip-On Flange NPS 4" S150#	4.00
Slip-On Flange NPS 2 1/2" S150#	1.00
Gate valve NPS 4" S150#	1.00
Check valve NPS 4" S150#	1.00
Concentric reducer 4" x 2 1/2"	1.00
Pipe NPS 2 1/2"	0.40
Reducing tee 6" x 4"	1.00

16.4 Step 1: Calculating Piping System Weight

The first step is to calculate the total weight of the components. To do this, the bill of materials and technical reference tables are used to obtain the weight of the suction and discharge lines, detailed in Table 16.3.

t

Result: The total weight of the piping system is **327.09 kg**.

Table 16.3. List and Weight of Materials for this Project

NPS	Material	Unit Weight (kg)	Quantity	Total Weight (kg)	Reference Table
4"	90° LR Elbow SCH 40	3.86	3.00	11.58	Table 5.1
4"	A 53 Pipe SCH 40	16.06	4.00	64.24	Table 4.1
3"	A 53 Pipe SCH 40	11.30	0.50	5.65	Table 4.1
2 1/2"	A 53 Pipe SCH 40	8.50	0.50	4.25	Table 4.1
4"	SO Flange S150#	5.44	6.00	32.64	Table 12.1
3"	SO Flange S150#	4.08	1.00	4.08	Table 12.1
2 1/2"	SO Flange S150#	3.63	1.00	3.63	Table 12.1
4"	Gate Valve S150#	64.00	2.00	128.00	Table 8.1
4"	Y-Strainer	40.00	1.00	40.00	—
6" x 4"	Reducing Tee 6" x 4"	15.00	2.00	30.00	—
4" x 3"	Eccentric Reducer 4" x 3"	1.59	1.00	1.59	—
4" x 2 1/2"	Concentric Reducer 4" x 2 1/2"	1.43	1.00	1.43	—
			Total Weight (kg):	**327.09**	

16.5 Step 2: Calculating Mhr for Supports (Fabrication and Assembly)

To use the standard productivity rates from Table 9.1, which are expressed in Mhr/m, the total system weight is first converted into an **equivalent length**.

16.5.1 Calculating Equivalent Length

This value is obtained by dividing the total system weight by the unit weight of a reference pipe (in this case, NPS 4" SCH 40), using the following formula:

$$\text{Equivalent Length} = \frac{327.09\,\text{kg}}{16.06\,\text{kg/m}} = \mathbf{20.37}\,\text{m}$$

16.5.2 Estimating Mhr for Supports

$$\text{Total Mhr} = 20.37\,\text{m} \times 1.17\,\text{Mhr/m} = \mathbf{23.83}\,\text{Mhr}$$

16.6 Step 3: Mhr for Piping Blasting and Painting

16.6.1 Calculating Surface Area

The first step is to determine the total surface area. For this, a **calculation length** is established by summing the straight pipe runs and adding one linear meter for each fitting, resulting in 23 meters for this case.

The total area is obtained by multiplying this length by the unit area of the reference pipe:

$$\text{Total Area} = \text{Calculation Length} \times \text{Unit Area}$$

$$23\,\text{m} \times 0.36\,\text{m}^2/\text{m} = \mathbf{8.28}\,\mathbf{m^2}$$

16.6.2 Estimating Man-hour

Once the surface area is calculated, the man-hours are estimated by applying standard productivity rates for blasting and painting. Table 16.4 details this calculation, for which the same conservative values from the previous example were used.

Table 16.4. Man-Hour Estimation for Abrasive Blasting and Painting

Activity	**Mhr per m^2**	**Surface Area to be Treated (m^2)**	**Total (Mhr)**	**Reference Table**
Abrasive Blasting	1.50	8.28	12.42	—
Painting and touch-up	2.20	8.28	18.22	—

16.7 Step 4: Mhr for Support Blasting and Painting

Unlike piping, the surface area of supports is estimated indirectly from their weight. The process consists of two stages.

16.7.1 Calculating Weight and Surface Area

First, the total weight of the supports is determined by multiplying the system's equivalent length by the corresponding ratio from Table 9.1 (2 kg/m):

$$20.37\,\text{m} \times 2\,\text{kg/m} = 40.74\,\text{kg} \approx 41\,\text{kg}$$

Next, the weight is converted to surface area using the standard coefficient for light structures (35 m^2/ton):

$$0.041\,\text{ton} \times 35\,\text{m}^2/\text{ton} = 1.44\,\text{m}^2$$

16.7.2 Estimating Man-hours

With the surface area calculated, the man-hours required for the anti-corrosion treatment can be estimated, as detailed in the following table:

Table 16.5. Calculation of Total Man Hours for Abrasive Blasting and Painting

Activity	Kg of Support (A)	m^2/Ton (B)	Surface Area (A × B = C)	Mhr/m^2 (D)	Total (C × D = E)	Reference Table
Abr. Blasting	41.00	35.00	1.44	1.50	2.16	—
Painting	41.00	35.00	1.44	2.20	3.17	—
				Total Mhr:	**5.33**	

16.8 Analysis of Execution Scenarios

16.8.1 Case A: Full On-Site Assembly and Welding

This scenario represents the traditional execution approach, where all piping components (pipes, elbows, flanges) are transported to the site to be assembled and welded in their final position. This method is highly dependent on field conditions, such as weather, access availability, and work permit management.

Productivity is typically lower compared to workshop prefabrication, as the crew must work in a less controlled environment. Below, the man-hours required for this case are calculated using the standard productivity rates for on-site work detailed in Table 16.6.

Table 16.6. Man-Hour Calculation for On-Site Assembly and Welding

NPS	Material	Quantity	Mhr	Total Mhr	Reference Table
4"	90° LR Elbow	3.00	5.00	15.00	Table 5.1
4"	Weld nipples	2.00	Included in fittings	N/A	—
4"	Valve, flanged gate S150#	2.00	2.00	4.00	Table 8.1
4"	Y-Strainer	1.00	2.00	2.00	—
4"	SO Flange S150#	6.00	2.90	17.40	Table 12.1
3"	SO Flange S150#	1.00	2.48	2.48	Table 12.1
2 1/2"	SO Flange S150#	1.00	2.00	2.00	Table 12.1
6" x 4"	Reducing tee	2.00	9.50	19.00	Table 6.1
4" x 3"	Eccentric reducer	1.00	4.50	4.50	Table 6.1
4" x 2 1/2"	Concentric reducer	1.00	4.20	4.20	Table 6.1
			Total Mhr:	**70.58**	

16.9 On-Site Productivity Analysis

To validate the estimate, the daily productivity of a typical assembly crew—composed of **one welder, one pipefitter, and one helper**—is analyzed. This crew generates a total of 24 man-hours (Mhr) per workday.

Considering that the total scope is 99 inches to be welded and that assembly consumes 64.58 Mhr (excluding quick-install items like valves and the strainer), the expected crew productivity can be calculated.

16.9.1 Daily Productivity Calculation

The formula relates the work to be done (inches) with the effort required (Mhr):

$$\text{Productivity} = \frac{\text{Total Inches} \times \text{Mhr per Workday}}{\text{Total Mhr Consumed}}$$

$$\text{Productivity} = \frac{99'' \times 24\,\text{Mhr/workday}}{64.58\,\text{Mhr}} = \mathbf{36.80}\,\text{inches/workday}$$

This productivity rate of 36.80 inches per day is a realistic value validated by the author's experience in comparable projects, which confirms that the hour estimate is appropriate for on-site work conditions.

16.9.2 Case B: Assembly with Workshop Prefabrication

This scenario analyzes the prefabrication strategy, where most joints are welded in a workshop. This method significantly reduces field work time thanks to controlled conditions that improve efficiency, quality, and safety.

For the man-hour estimation, a **50% reduction** in execution time is assumed for each joint welded in the workshop compared to its on-site equivalent. Therefore, fieldwork is minimized, limited to the essential welds for final assembly and connection of the system.

In this project, the joints that must be executed on-site are:

➔ The connection of the two 6"x4" tees to the existing headers.

➔ The final tie-in welds to connect the prefabricated spools.

The following table breaks down the total Mhr calculation for this scenario, combining the reduced cost of prefabricated joints with that of the critical joints executed on-site.

Table 16.7. Man-Hour Calculation with Shop Prefabrication

NPS	Item	Quantity	Mhr	Total Mhr	Reference Table
4"	Shop-fabricated joints	12.00	1.25 (2.50/2)	15.00	Table 5.1
3"	Shop-fabricated joints	2.00	1.00 (2.00/2)	2.00	Table 5.1
2 1/2"	Shop-fabricated joints	2.00	0.88 (1.75/2)	1.76	Table 5.1
4"	Field joints	4.00	2.50	10.00	Table 5.1
6"	Field joints	4.00	3.50	14.00	Table 5.1
4"	Gate Valve NPS 4" S 150#	2.00	2.00	4.00	Table 8.1
4"	Y-Strainer	1.00	2.00	2.00	—
			Total Mhr:	**48.76**	

16.10 Final Scenario Comparison

Workshop prefabrication (Case B) offers substantial advantages over traditional on-site assembly (Case A), though it also introduces a critical risk that must be managed. The pros and cons of prefabrication are detailed below:

- ➔ **Higher Quality and Precision:** The controlled environment of a workshop allows for strict dimensional control, higher integrity welds, and efficient execution of non-destructive testing (NDT). This minimizes the need for costly corrections during field assembly.
- ➔ **Efficiency and Cost Reduction:** Labor utilization is optimized, production becomes independent of weather conditions, and final assembly is accelerated. As a result, overall project timelines and associated indirect costs are reduced.
- ➔ **Safety and Cleanliness:** Prefabrication promotes a safer, more orderly, and cleaner work environment—a key factor in highly regulated industries such as pharmaceuticals or food and beverage.

Critical Risk to Consider. Despite its benefits, prefabrication presents a key risk: **dimensional errors**. A piece fabricated incorrectly in the workshop, whether due to an engineering or manufacturing flaw, can trigger a chain of costly corrections throughout the entire line, potentially nullifying the savings achieved. Therefore, precise detail engineering and fluid communication with the workshop are fundamental.

Table 16.8 summarizes the man-hours.

Table 16.8. Comparative Summary of Total Man-Hours

Description	Mhr Case II A	Mhr Case II B
Piping assembly and welding	70.58	48.76
Support fabrication and assembly	23.83	23.83
Piping painting	30.64	30.64
Support painting	5.33	5.33
Total Mhr	130.38	108.56

16.10.1 Estimate Adjustment: Correction Factors

Once the base man-hour estimate is obtained, it is crucial to adjust it according to the specific project conditions. These **correction factors** quantify the impact of variables that can affect the actual productivity of the workforce. The key factors for this case are analyzed below.

16.10.2 Correction A: Project-Specific Variable Analysis

The impact of various conditions is evaluated to determine if an adjustment to the base estimate is required. A factor of 1.00 indicates that conditions are standard and do not affect productivity.

Table 16.9. Evaluation of Specific Variables

Factor Analyzed	Case Analysis	Factor
Team Experience	The task is simple and routine for a team with the necessary qualifications. No cost overruns are anticipated due to a lack of experience.	1.00
Engineering Quality	As this is a minor modification, the documentation (drawings and specifications) is considered accurate and complete.	1.00
Labor and Union Environment	The project does not present severe demands. Union influence for work of this magnitude is insignificant.	1.00
Site Conditions	The work is performed in a covered area of the plant, with a stable concrete floor and easy access for personnel and equipment.	1.00
Climatic Factors	The work area is an enclosed interior, so the team is protected from any adverse weather conditions.	1.00
Work Height	The tasks are executed at a standard height, with no performance losses associated with work at height anticipated.	1.00

Conclusion of Analysis A: The project conditions are standard, and no deviations are anticipated. **Correction factor A is 1.00**.

B. Correction for Congestion & Restricted Access (Brownfield)

This factor quantifies the productivity differential between new construction on open ground (*Greenfield*) and revamp works inside an operating industrial plant (*Brownfield*). Executing modifications or *tie-ins* on live process lines and in-service equipment imposes safety restrictions, permit-related idle time, and logistical constraints that directly degrade craft crew efficiency.

Applied Correction: A $10\,\%$ increase is applied to baseline direct man-hours. Correction factor B is **1.10** $(1.10\times)$.

16.10.3 Correction C: Material and WPS Factor

This factor is applied if the materials differ from the standard carbon steel used in the tables. In this case, the specified materials are carbon steel, so no adjustment is required.

Applied Correction: No correction is required. **Correction factor C is 1.00**.

16.10.4 Calculation of the Final Adjusted Estimate

To obtain the final man-hour budget, the base estimate is multiplied by each of the defined correction factors (A, B, and C):

$$\text{Total Factor} = A \times B \times C = 1.00 \times 1.10 \times 1.00 = \mathbf{1.10}$$

The final adjusted values are shown in the summary table.

Table 16.10. Final Adjusted Estimate Calculation

Description	**Mhr Case A (On-Site Welding)**	**Mhr Case B (With Prefabrication)**
Total Mhr (Base)	130.38	108.56
Additional for site location (10%)	13.04	10.86
Final Estimated Total	**143.42**	**119.42**

Abbreviations and Units

STH = Standard Table Hours.

Mhr = Man-hour.

Mhr/unit = Man-hours per unit.

kg/m = kilograms per meter.

g/joint = grams per joint.

mm = millimeter = 0.03937 inch.

m = meter = 3.281 feet.

km = kilometer = 0.6214 miles.

m^2 = square meter = 10.764 ft^2.

m^3 = cubic meter = 35.314 ft^3.

ft = linear foot = 0.3048 meters.

L = Liter = 61.023 in^3 = 0.03531 ft^3 = 0.2642 U.S. gal.

kg = kilogram = 2.2046 lb.

lb = pounds = 0.4536 kg.

t = metric ton = 1000 kg.

bar = 14.5 psi (lb/in^2).

NPS = Nominal Pipe Size in inches; OD = Outside Diameter; ID = Inside Diameter.

t = thickness; V = volume.

max. = maximum; min. = minimum.

kW = kilowatt = 1.341 horsepower; kWh = kilowatt-hour.

kVA = kilovolt-ampere.

kg/Mhr = kilogram per Man-hour.

Mhr/m = Man-hours per meter.

k = correction factor.

XS = Extra Strong.

XXS = Double Extra Strong.

About the Author

Figure 16.3. Photo of the Author.

Why Trust This Manual?

The reliability of this manual is grounded in its origin: real and proven field experience. Its author, **Gustavo Miguel Cinca**, a Chemical Engineer who graduated with Honors, has dedicated his career to the execution of complex industrial projects, leading teams from site management to the presidency of a renowned assembly and construction company. His career spans from refineries and chemical plants to gas pipelines, both in Argentina and internationally.

The methodologies and performance rate tables presented here are not an academic exercise. They are the same tools Gustavo developed and used to successfully compete for, bid on, and win tenders.

Therefore, what you have in your hands is an eminently practical and realistic guide, forged through decades of direct, on-the-ground experience.

www.ingramcontent.com/pod-product-compliance
Ingram Content Group UK Ltd.
Pitfield, Milton Keynes, MK11 3LW, UK
UKHW051508290926
15471UKWH00008B/1491

9 786310 106885